**Springer Theses**

Recognizing Outstanding Ph.D. Research

## Aims and Scope

The series "Springer Theses" brings together a selection of the very best Ph.D. theses from around the world and across the physical sciences. Nominated and endorsed by two recognized specialists, each published volume has been selected for its scientific excellence and the high impact of its contents for the pertinent field of research. For greater accessibility to non-specialists, the published versions include an extended introduction, as well as a foreword by the student's supervisor explaining the special relevance of the work for the field. As a whole, the series will provide a valuable resource both for newcomers to the research fields described, and for other scientists seeking detailed background information on special questions. Finally, it provides an accredited documentation of the valuable contributions made by today's younger generation of scientists.

## Theses may be nominated for publication in this series by heads of department at internationally leading universities or institutes and should fulfill all of the following criteria

- They must be written in good English.
- The topic should fall within the confines of Chemistry, Physics, Earth Sciences, Engineering and related interdisciplinary fields such as Materials, Nanoscience, Chemical Engineering, Complex Systems and Biophysics.
- The work reported in the thesis must represent a significant scientific advance.
- If the thesis includes previously published material, permission to reproduce this must be gained from the respective copyright holder (a maximum 30% of the thesis should be a verbatim reproduction from the author's previous publications).
- They must have been examined and passed during the 12 months prior to nomination.
- Each thesis should include a foreword by the supervisor outlining the significance of its content.
- The theses should have a clearly defined structure including an introduction accessible to new PhD students and scientists not expert in the relevant field.

Indexed by zbMATH.

Fernando Vaquerizo Villar

# Automated Analysis of the Oximetry Signal to Simplify the Diagnosis of Pediatric Sleep Apnea

## From Feature-Engineering to Deep-Learning Approaches

Doctoral Thesis accepted by
University of Valladolid, Valladolid, Spain

Springer

*Author*
Dr. Fernando Vaquerizo Villar
Biomedical Engineering Group
E.T.S.I. de Telecomunicación
University of Valladolid
Valladolid, Spain

*Supervisors*
Dr. Roberto Hornero Sánchez
Biomedical Engineering Group
E.T.S.I. de Telecomunicación
University of Valladolid
Valladolid, Spain

Dr. Daniel Álvarez González
Biomedical Engineering Group
E.T.S.I. de Telecomunicación
University of Valladolid
Valladolid, Spain

ISSN 2190-5053 ISSN 2190-5061 (electronic)
Springer Theses
ISBN 978-3-031-32831-2 ISBN 978-3-031-32832-9 (eBook)
https://doi.org/10.1007/978-3-031-32832-9



This Springer imprint is published by the registered company Springer Nature Switzerland AG
The registered company address is: Gewerbestrasse 11, 6330 Cham, Switzerland

*To all of you who have supported me throughout these years*

# Supervisors' Foreword

It is our honor to introduce the thesis work of Dr. Fernando Vaquerizo-Villar, who carried out a thorough and outstanding research that lead to major contributions in the framework of biomedical engineering and, particularly, in the context of decision support systems for pediatric sleep-related breathing disorders. The path that leads to the thesis began in 2014, when Dr. Vaquerizo-Villar started his collaboration with our research group (Biomedical Engineering Group) at University of Valladolid, Spain. His first contact with biomedical signal processing was the studies that Dr. Vaquerizo-Villar carried out during his final degree project of the bachelor's degree in Specific Telecommunication Technologies Engineering ("Analysis of the blood oxygen saturation signal to assist in the diagnosis of pediatric sleep apnea-hypopnea syndrome", 2014) and his master's thesis in Telecommunications Engineering ("Feature extraction, selection, and classification of the oximetry signal to assist in the automated detection of pediatric sleep apnea-hypopnea syndrome", 2016), reaching the highest qualifications in both of them. Since then, he was endeavoring to gain deeper knowledge on biomedical data analysis through signal processing and artificial intelligence techniques. As a natural consequence, he embarked on his career as Ph.D. student in the Information and Telecommunications Technologies Doctoral Programme. In the period 2016–2021, Dr. Vaquerizo-Villar collaborated in several R&D projects, mainly focused on biomedical signal processing in the field of sleep medicine. He defended his doctoral thesis at University of Valladolid in December 2021 ("Automated analysis of the oximetry signal to simplify the diagnosis of pediatric sleep apnea: from feature-engineering to deep-learning approaches"), in which he obtained the highest qualification (*Summa Cum Laude*).

The thesis of Dr. Vaquerizo-Villar summarizes an original and outstanding research focused on applying novel signal processing algorithms in order to enhance the diagnostic ability of the oximetry signal in the framework of pediatric obstructive sleep apnea (OSA). The original journal papers included in the thesis present novel feature engineering and deep learning methodologies. On the one hand, three novel feature extraction algorithms (bispectrum, wavelet, and detrended fluctuation analysis) were used to provide complimentary features that help to further characterize apneic events linked to pediatric OSA. On the other hand, a deep learning methodology based on convolutional neural networks was employed to automatically estimate pediatric OSA severity from raw oximetry data. These novel approaches led to an enhanced characterization of the changes in the oximetry signal caused by apneic events in pediatric subjects, as well as to an increase in the diagnostic capability of nocturnal oximetry in the context of childhood OSA. We feel that this research could contribute to the use of clinical screening tools to diagnose pediatric OSA based on the automated analysis of the oximetry signal.

Valladolid, Spain
March 2023

Dr. Roberto Hornero Sánchez
Dr. Daniel Álvarez González

# Foreword by the Bioengineering Group of the Automatic Control Spanish Association (CEA)

Engineering and technological innovations over the past few decades have rapidly increased the quality and accessibility of health care worldwide. The promotion and translation of such broad and cutting-edge emergent technologies into healthcare practices are more and more encouraged for the quality of life globally. For this purpose, leaders in bioengineering research need to work together with clinicians and healthcare industry leaders to develop novel diagnostics and treatments for patient populations. The work presented in this monograph written by Dr. Fernando Vaquerizo is a contribution in the area of bioengineering, which includes significant improvements in signal processing algorithms and deep learning methodologies to enhance the diagnosis of pediatric obstructive sleep apnea, a high prevalent respiratory disorder among children. Thus, this work includes the investigation of novel feature engineering and deep learning approaches that could be applied to obtain complementary information from oximetry signals for the improvement of its diagnostic ability.

The contents of this monograph are focused on the use of frequency domain (bispectrum and wavelet analysis) and nonlinear (detrended fluctuation analysis) analysis techniques to provide discriminative features from the oximetry signal for the first time applied in the context of the above-mentioned childhood sleep condition. Moreover, high-performance pattern recognition models have been tested including multiclass classification, regression, and novel deep learning models for automated extraction of features with higher accuracy compared to state-of-the-art classification methodologies. Special focus is given to the development of a diagnostic protocol that is derived from the proposed deep learning modeling approach, toward a diagnostic testing with higher usability and less obstructiveness.

The monograph is based on the first author's doctoral thesis, awarded for best Ph.D. thesis in bioengineering during the 2020 edition of the award call organized by the Bioengineering Group of the Spanish Committee of Automatic Control (*Comité Español de Automática*, CEA). This annual award is aimed at recognizing the outstanding Ph.D. research in the field of bioengineering. Participation requires at least one of the theses' supervisors to be a partner of CEA and a member of the

Bioengineering Group. The jury is composed of three well-known doctors in the field: Two of them are partners of CEA, and the third one is a foreign professor.

A total number of 4 Ph.D. theses were submitted to this edition in 2020. Notably, their authors show altogether a scientific production of about 45 publications in international indexed journals, with more than 20 works published in the top-quartile journals of their categories.

Fernando Vaquerizos's Ph.D. thesis was selected by the Bioengineering Group of CEA as the best among other excellent candidates. Through this publication, we hope that Fernando's work reaches a large international audience and becomes a valuable source of information and inspiration for other researchers and students in this field which holds tremendous promise for advancing human health and well-being. We have no doubt that Fernando's work will continue to inspire and shape the field of bioengineering, and I am confident that his contributions will have a lasting impact on the scientific community and society as a whole.

Once again, congratulations to Fernando on this exceptional achievement, and all the best in your future endeavors.

Dr. Juan C. Moreno
Spanish National Research Council
Cajal Institute
Neural Rehabilitation Group
Madrid, Spain
http://www.neuralrehabilitation.org

**Parts of this thesis have been published in the following articles**

**Journals**

1. **Fernando Vaquerizo-Villar**, Daniel Álvarez, Leila Kheirandish-Gozal, Gonzalo C. Gutiérrez-Tobal, Verónica Barroso-García, Eduardo Santamaría-Vázquez, Félix del Campo, David Gozal, Roberto Hornero, "A convolutional neural network architecture to enhance oximetry ability to diagnose pediatric obstructive sleep apnea", *IEEE Journal of Biomedical and Health Informatics*, vol. 25 (8), August, 2021, DOI: 10.1109/JBHI.2020.3048901. Impact factor in 2021: 7.021, D1 in "MATHEMATICAL & COMPUTATIONAL BIOLOGY" (JCR-WOS).
2. **Fernando Vaquerizo-Villar**, Daniel Álvarez, Leila Kheirandish-Gozal, Gonzalo C. Gutiérrez-Tobal, Verónica Barroso-García, Andrea Crespo, Félix del Campo, David Gozal, Roberto Hornero, "Wavelet analysis of oximetry recordings to assist in the automated detection of moderateto-severe pediatric sleep apnea-hypopnea syndrome", *PLOS One*, vol. 120, pp. 155–166, December, 2018, DOI: 10.1371/journal.pone.0208502. Impact factor in 2018: 2.776, Q2 in "MULTIDISCIPLINARY SCIENCES" (JCR-WOS).
3. **Fernando Vaquerizo-Villar**, Daniel Álvarez, Leila Kheirandish-Gozal, Gonzalo C. Gutiérrez-Tobal, Verónica Barroso-García, Andrea Crespo, Félix del Campo, David Gozal, Roberto Hornero, "Detrended fluctuation analysis of the oximetry signal to assist in paediatric sleep apnoea–hypopnoea syndrome diagnosis", *Physiological Measurement*, vol. 39(11), p. 114006, November, 2018, DOI: 10.1088/1361-6579/aae66a. Impact factor in 2018: 2.246, Q3 in "ENGINEERING, BIOMEDICAL" (JCR-WOS).
4. **Fernando Vaquerizo-Villar**, Daniel Álvarez, Leila Kheirandish-Gozal, Gonzalo C. Gutiérrez-Tobal, Verónica Barroso-García, Andrea Crespo, Félix del Campo, David Gozal, Roberto Hornero, "Utility of bispectrum in the screening of pediatric sleep apnea-hypopnea syndrome using oximetry recordings", *Computer Methods and Programs in Biomedicine*, vol. 156, p. 141–149, March, 2018, DOI: 10.1016/j.cmpb.2017.12.020. Impact factor in 2018: 3.424, Q1 in "COMPUTER SCIENCE, THEORY & METHODS" (JCR-WOS).

Parts of this thesis have been published in the following articles:

[illegible]

1. Fernando [illegible]
2. Fernando [illegible]
3. Fernando [illegible]
4. Fernando [illegible]

# Acknowledgements

First, I would like to thank my Ph.D. advisors, Dr. Roberto Hornero Sánchez and Dr. Daniel Álvarez González, for supervising me during all these years in my research activity. Since they began to supervise my bachelor's degree project in 2014, they have trusted me, showing an excellent human level at all times.

I also want to extend my gratitude to all the current and former members of the Biomedical Engineering Group (*Grupo de Ingeniería Biomédica*, GIB): Jesús, Carlos, María, Gonzalo, Javier G., Rebeca, Alejandro, Luis F., Víctor M., Pablo, Saúl, Verónica, Roberto R., Eduardo, Jorge, Adrián, Víctor R., Víctor G., Aarón, Marcos, Sergio, Diego, Selene, Clara, Enrique, Robert C. C., Javier O., and Celia. Apart from being an excellent research group to develop the doctoral thesis, GIB offers a great work environment, with interesting methodological discussions, as well as unforgettable jokes and anecdotes in the coffee breaks and happy meals. Some of them even ended up on the covers of the best magazines. I am especially grateful to Gonzalo and Verónica, who have greatly contributed to my training by reviewing each of my articles and conference papers. Nor can I forget Jesús, Carlos, and María, who have helped me enormously in the teaching activities that I have carried out during these years.

I want to extend my gratitude to the Dr. David Gozal and the Dr. Leila Kheirandish Gozal, from the University of Missouri School of Medicine, as well to as the Dr. Félix del Campo from the *Hospital Universitario Río Hortega*. Their medical knowledge and extensive collaboration have been essential to carry out this thesis.

I would also like to thank Dr. Thomas Penzel and the members from the Interdisciplinary Center of Sleep Medicine from the *Charité UniversitätsMedizin Berlin* and the Cardiovascular Physics Group from the Humboldt University in Berlin for accepting me in one of the best sleep research centers in the world. My research stay in Berlin was an enriching experience, which helped me to grow both personally and professionally. Thanks Thomas for giving me the opportunity to collaborate in international research projects. I am also very grateful to Niels Wessel, Jan Kraemer, and Hua Qin for their help, advice, and feedback with my work, as well as to all the nice people in Berlin who made me feel like home.

Finally, I would like to thank my family, who were the first to encourage me to follow this path. I have had the best parents, siblings, grandparents, uncles, cousins, and other family. You have worked hard since the day I was born to help me get here, although some of you will not be able to see it. Nor can I forget my girlfriend, nor my friends, who have become a second family, for their concern, support, and help to make the difficult look easy.

# Contents

# Acronyms

| | |
|---|---|
| AAP | American Academy of Pediatrics |
| AASM | American Academy of Sleep Medicine |
| Acc | Accuracy |
| AF | Airflow |
| AHI | Apnea-hypopnea index |
| ApEn | Approximate entropy |
| AUC | Area under the ROC curve |
| BI | Band of interest |
| BMI | Body mass index |
| BN | Batch normalization |
| BO-TPE | Bayesian optimization with tree-structured Parzen estimator |
| BUH | Burgos University Hospital |
| CHAT | Childhood Adenotonsillectomy Trial |
| CNN | Convolutional neural networks |
| CTM | Central tendency measure |
| CWT | Continuous wavelet transform |
| DFA | Detrended fluctuation analysis |
| DWT | Discrete wavelet transform |
| ECG | Electrocardiogram |
| EEG | Electroencephalogram |
| EMG | Electromyogram |
| EOG | Electrooculogram |
| FCBF | Fast correlation-based filter |
| FN | False negatives |
| FP | False positives |
| HHb | Deoxymehoglobin |
| HR | Heart rate |
| ICC | Intra-class correlation coefficient |
| JCR | Journal Citation Reports |
| Kappa | Cohen's kappa |
| LDA | Linear discriminant analysis |

| | |
|---|---|
| LR | Logistic regression |
| LR– | Negative likelihood ratio |
| LR+ | Positive likelihood ratio |
| LZC | Lempel-Ziv complexity |
| MLP | Multi-layer perceptron |
| MOS | McGill oximetry score |
| MSE | Multiscale entropy |
| NPV | Negative predictive value |
| $O_2Hb$ | Oxyhemoglobin |
| ODI | Oxygen desaturation index |
| ODI3 | 3% Oxygen desaturation index |
| ODI4 | 4% Oxygen desaturation index |
| OSA | Obstructive sleep apnea |
| POW | Pulse oximetry watch |
| PPG | Photoplethysmography |
| PPV | Positive predictive value |
| PSD | Power spectral density |
| PSG | Polysomnography |
| QDA | Quadratic discriminant analysis |
| ReLU | Rectified linear unit |
| RMSE | Root mean square error |
| ROC | Receiver-Operating Characteristics |
| RP | Respiratory polygraphy |
| SAHS | Sleep apnea-hypopnea syndrome |
| SampEn | Sample entropy |
| Se | Sensitivity |
| Sp | Specificity |
| $SpO_2$ | Blood oxygen saturation |
| STFT | Short time Fourier transform |
| SU | Symmetrical uncertainty |
| SVM | Support vector machine |
| TN | True negatives |
| TP | True positives |
| UofC | University of Chicago |
| WT | Wavelet transform |

# Chapter 1
# Introduction

Obstructive sleep apnea (OSA) is a high prevalent respiratory disorder in the pediatric population. Untreated pediatric OSA is associated with significant adverse consequences affecting metabolic, cardiovascular, neurocognitive, and behavioral systems, thus resulting in a decline of overall health and quality of life. Consequently, it is of paramount importance to accelerate the diagnosis and treatment in these children.

Overnight polysomnography (PSG) is the gold standard to diagnose OSA in children. This test requires an overnight stay of pediatric subjects in a specialized sleep laboratory, as well as the recording of up to 32 biomedical signals. These recordings are used to quantify respiratory events in order to obtain the apnea-hyponea index (AHI), which is used to establish pediatric OSA severity. Nonetheless, PSG is technically complex, time-consuming, costly, highly intrusive for the children, and relatively unavailable, thus delaying the access for both the diagnosis and treatment. Consequently, simplified diagnostic techniques become necessary.

In an effort to overcome these drawbacks and increase the accessibility of pediatric OSA diagnosis, many simplified alternative procedures have been developed. Among these, a common approach is the analysis of the blood oxygen saturation ($SpO_2$) signal from overnight oximetry due to its easy acquisition and interpretation, as well as its suitability for children. Many studies have demonstrated the utility of the automated analysis of $SpO_2$ recordings to help in adult OSA diagnosis. Conversely, the preceding studies focused on pediatric patients reported lower accuracies than those reached in the case of adults, suggesting the need to seek novel signal processing algorithms that provide additional information from the $SpO_2$ signal for the particularities of childhood OSA.

In this context, the present doctoral thesis aims to design, develop, and assess novel feature extraction and deep-learning methodologies to improve the diagnosis ability of the oximetry signal in the context of pediatric OSA.

In this chapter, the general context of biomedical signal processing and deep learning is firstly and concisely described in Sect. 1.1. Section 1.2 provides a description of pediatric OSA, including its risks and adverse consequences. Section 1.3 focuses



F. Vaquerizo Villar, *Automated Analysis of the Oximetry Signal to Simplify the Diagnosis of Pediatric Sleep Apnea*, Springer Theses,
https://doi.org/10.1007/978-3-031-32832-9_1

on the gold standard for pediatric OSA diagnosis, nocturnal PSG, and its limitations. Finally, Sect. 1.4 is devoted to explain oximetry as an alternative to PSG for the diagnosis of pediatric OSA and Sect. 1.5 provides a description of state-of-the-art studies focused on the analysis of the oximetry signal as a simplified tool in the diagnosis of pediatric OSA.

## 1.1 The Biomedical Signal Processing Framework: Feature-Engineering and Deep Learning

Biomedical signals convey information on the functioning of the human body [1]. The study of these signals allows to analyze the properties of the underlying biological systems [1], which makes possible to identify several pathological conditions [2]. Nonetheless, the physiological information contained in these signals cannot be typically extracted in a visual way [2]. In this respect, biomedical signal processing provides methods that help to understand and characterize the hidden information from these signals that can not be obtained through visual assessment [2]. It also allows to develop automated systems for the diagnosis, treatment, and/or monitoring of a wide range of pathologies [2].

In this type of systems (e.g., the screening of pediatric OSA), the automated analysis of biomedical signals has been traditionally performed following a feature-engineering methodology [3], which consists of three main stages. In the first stage, known as feature extraction, the hidden characteristic information (features) about the biomedical signals is obtained [3, 4]. To extract these features, different algorithms based on mathematical methods are used such as statistical, morphological, frequency domain, time-frequency, or nonlinear analysis [3–5]. The second stage is the use of automatic feature selection methods to find the relevant and non-redundant information among that extracted in the previous step [5]. Finally, in the third stage, the selected information is used to train pattern recognition algorithms in order to obtain predicted models aimed at providing a diagnostic decision [3, 5]. The range of patter-recognition methods include from weak classifiers like logistic regression (LR) and Fisher linear discriminant analysis (LDA) to more complex algorithms such as support vector machines (SVM), Bayesian classifiers, and multi-layer perceptron (MLP) neural networks [3, 5].

Deep learning has emerged in recent year as a novel methodological approach aimed at changing the paradigm of data processing [6]. Conventional feature-engineering approaches have two main disadvantages: (*i*) a human expert must determine which relevant features obtain from the input data, and (*ii*) these methods provide a low level of abstraction that limits their capability to learn complex features from the data. These issues may result in missing important information from the data. In contrast to conventional approaches, a deep-learning model automatically discovers the intricate information in the data [6]. In this regard, deep-learning methods automatically learn complex patterns and extract features from raw data by the use of various processing layers with multiple levels of representation [6]. Starting from the representation at the lowest level, the raw input data, simple non-linear modules transform them into representations at a higher abstract level [6]. With an architecture composed of enough transformations, deep-learning algorithms can learn very complex features from the data. As aforementioned, these algorithms have improved predictive performances in a broad range of traditionally challenging domains, such as image, genomics, and signal processing [6]. Specifically, in the biomedical signal processing field, these algorithms have beaten conventional methods in many relevant domains, including sleep stage scoring [7], congestive heart failure diagnosis [8], epileptic seizure detection [9], and brain-machine interfaces [9].

This Doctoral Thesis is aimed at enhancing the diagnosis ability of the oximetry signal in the context of childhood OSA. For this purpose, novel feature-engineering and deep-learning methodologies have been developed and assessed.

## 1.2 Pediatric Obstructive Sleep Apnea (OSA)

Although originally described for adults, OSA has been recognized in recent years as a high prevalent condition among children (1.2–5.7%), with etiological, diagnostical, and therapeutical considerations that are different for the pediatric population [10, 11]. According to the American Academy of Pediatrics (AAP), childhood OSA is a respiratory disorder marked by repetitive episodes of complete absence (apnea) and/or considerable reduction (hypopnea) of airflow during sleep [11]. It is associated with the presence of nocturnal symptoms that cause disturbed sleep. Thus, apneic (apneas and hypopneas) events derive in inadequate gas exchange, leading to hypercapnia and hypoxia states, which induce oxygen desaturations, arousals, and sleep fragmentation. Gasping and snoring also occur frequently [12].

As a consequence of these symptoms, OSA have many negative effects that reduce health and quality of life of the children [10, 11]. In this respect, children suffering from OSA are at an increased risk for developing cardiovascular morbidities, such as systemic hypertension, changes in blood pressure regulation, and altered left ventricular geometry [10]. OSA during childhood may also lead to neurobehavioral abnormalities, such as cognitive deficits, reduced academic achievements, hyperactivity, aggressive behavior, and excessive daytime sleepiness [13]. Moreover, it is also related to the metabolic syndrome, which includes hypertension, insulin resistance, dyslipidemia, and obesity [10]. Finally, somatic growth impairment has been related to pediatric OSA as well [14].

Despite its high prevalence, pediatric OSA is an underdiagnosed condition [15]. Estimations indicate that approximately 90% of the affected children have not been diagnosed yet [15]. Treatment interventions for pediatric OSA have led to a reduction in neurocognitive, cardiometabolic, and growth stunting risks [16]. Nonetheless, the low percentage of diagnosis, together with the high prevalence, result in a high number of children being exposed to its adverse consequences.

## 1.3 Pediatric OSA Diagnosis: Polysomnography (PSG)

OSA is diagnosed by means of the overnight PSG test, which acts as "gold standard" [11]. During PSG, multiple neurophysiological and cardiorespiratory signals from patients are monitored and recorded: electroencephalogram (EEG), electromyogram (EMG), electrooculogram (EOG), electrocardiogram (ECG), oronasal airflow (AF), abdominal and chest wall movements, respiratory effort, $SpO_2$, and photoplethysmography (PPG), among others [17]. Thus, patients need to stay a complete night in a sleep laboratory, where skilled staff care for them as well as monitor the course of the test. After the PSG, the sleep recordings need an offline inspection to annotate apneas and hypopneas in order to compute the AHI, which is the clinical variable employed to establish a diagnosis [17].

Despite the well-known effectiveness of PSG, it presents several limitations. PSG is a complex test due to the necessity to record a high number of signals, which requires that patients spend at least one night in a sleep laboratory [18]. In addition, trained personnel is needed to be responsible for the children and a proper development of the test, resulting in high hospital expenses [18]. Similarly, apneic events are manually scored by trained specialists, which is labor intensive and may result in subjective diagnoses [18]. Finally, the nature of the PSG requires to perform the test out of the sleep environment of the patients and with the use of multiple sensors placed on their bodies, which results highly uncomfortable and intrusive for children [19]. This may derive in obtaining sleep recordings which are not representative of natural sleep, thus resulting in the need to repeat the diagnostic test [19].

Due to the complexity, cost, and time needed to analyze the sleep signals, available resources are not enough to cope with the high demand of OSA diagnosis [18]. This results in long waiting lists, thus hindering the diagnosis and treatment of the affected children [20]. These drawbacks, together with the high prevalence rate of pediatric OSA, have led the scientific community to explore the use of simplified screening tests [11, 21]. In this sense, the guidelines of the AAP recommend conducting alternative tests to address PSG unavailability, while still requiring more conclusive evidences about the efficacy of these tests [11].

## 1.4 Alternatives to PSG

In order to address the above-mentioned limitations, the use of portable monitoring equipment has been suggested as the main alternative to PSG in the diagnosis of pediatric OSA [11, 21]. According to the Portable Monitoring Task Force of the American Academy of Sleep Medicine (AASM), the equipment used in sleep studies can be classified into four types, depending on the number and type of the recorded signals [22]:

I. **Type I: Standard PSG**. This first type consists of the conventional PSG equipment, which requires the supervision of trained personnel in the hospital facilities.

These devices are considered the gold standard to which the remaining types must be compared.

II. **Type II: Comprehensive portable PSG**. These devices record a minimum of seven channels, including EOG, chin EMG, EEG, airflow, ECG or heart rate (HR), respiratory effort, and $SpO_2$. These studies do not require the presence of trained personnel. These devices allow to identify sleep stages and calculate the AHI.

III. **Type III: Modified portable sleep apnea testing**. These studies, also called respiratory polygraphy (RP) studies, include the recording of ventilation (a minimum of two respiratory movement signals or one respiratory movement signal and airflow), ECG or HR, and $SpO_2$.

IV. **Type IV: Continuous single-bioparameter or dual-bioparameter recording**. These devices only record one or two physiological signals, being oximetry traditionally one of these measurements. In addition, all the equipment that does not meet Type III criteria is included in this group.

### *1.4.1 Overnight Oximetry*

In recent years, there has been an increased interest in overnight oximetry as the main alternative to PSG in the context of pediatric OSA diagnosis due to its simplicity, reliability, and suitability for children [23]. Overnight oximetry records the $SpO_2$ signal in a non-invasive way with a pulse oximeter, usually located on the finger, toe, or earlobe of the patient [24].

$SpO_2$ expresses the amount of oxygen combined with the hemoglobin with respect to the total hemoglobin in the blood, the oxyhemoglobin ($O_2Hb$), which is responsible for transporting the blood oxygen to the tissues. The operating principle of the $SpO_2$ sensors is based on the optical properties of the hemoglobin [25]. $O_2Hb$ absorbs more infrared light, acquiring a red hue. On the contrary, deoxyhemoglobin (HHb) absorbs a higher amount of red light, thus having a more bluish hue. Pulse oximeters exploit this difference in the light absorption of $O_2Hb$ and HHb to obtain $SpO_2$. To achieve this, pulse oximeters contain two light-emitting diodes on one side of the finger that transmit at red (around 660 nm) and near infrared wavelengths (around 940 nm). On the opposite side of the finger, a photodiode is placed to measure the amount of red and infrared light absorbed by the tissues, which allows to determine $SpO_2$ [25].

Due to these easy acquisition of the $SpO_2$ signal, commercial pulse oximetry devices have been developed, which facilitate to perform the test in an unsupervised way at children's home [20, 26]. Apneic events from OSA induce recurrent decreases in the $SpO_2$ [27], also called oxygen desaturations, so that this signal contains important OSA-related information. Figure 1.1 shows the $SpO_2$ signal corresponding to (a) a no OSA pediatric subject (AHI $< 1$ events per hour), (b) a doubtful no OSA pediatric subject (AHI $< 1$ events per hour), (c) a mild OSA pediatric subject ($1 \leq$ AHI $< 5$ events per hour), (d) a doubtful mild OSA pediatric subject ($1 \leq$ AHI $< 5$ events per hour), (e) a moderate OSA pediatric subject ($5 \leq$ AHI $< 10$ events per hour), (f)

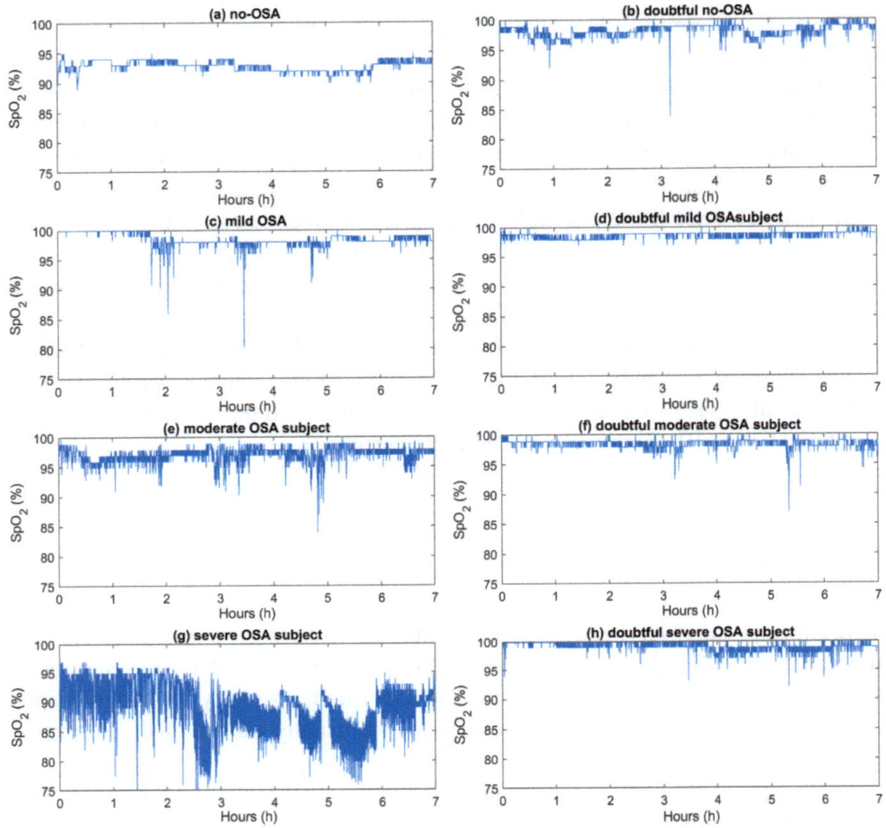


**Fig. 1.1** Examples of $SpO_2$ corresponding to: **a** a no OSA subject, **b** a doubtful no OSA subject, **c** a mild OSA subject, **d** a doubtful mild OSA subject, **e** a moderate OSA subject, **f** a doubtful moderate OSA subject, **g** a severe OSA subject, **h** a doubtful severe OSA subject. In doubtful subjects, it is difficult to visually discriminate the OSA severity group from the overnight $SpO_2$ profile, leading to the need of non-subjective automated analyses

a doubtful moderate OSA pediatric subject ($5 \leq$ AHI $< 10$ events per hour), (g) a severe OSA pediatric subject (AHI $\geq 10$ events per hour), and (h) a doubtful severe OSA pediatric subject (AHI $\geq 10$ events per hour). It can be observed that there are more oxygen desaturations as the severity of OSA increases. However, it is difficult to visually discriminate the $SpO_2$ signal from doubtful subjects.

## 1.5 State-of-the-art: Automated Analysis of the Oximetry Signal to Diagnose Pediatric OSA

The usefulness of $SpO_2$ recordings from nocturnal oximetry to help in the screening of pediatric OSA has been widely assessed, especially in the last years [23]. The analysis of this signal has been addressed by using conventional oximetric indices, as well as automated signal processing methodologies.

### *1.5.1 Conventional Oximetric Indices*

Previous studies predominantly used conventional oximetric indices for this task [20, 28–35]. In this respect, researchers mainly assessed the screening ability of oxygen desaturation index (ODI), which accounts for the number of drops of the $SpO_2$ signal larger than a defined threshold [30–32]. Kirk et al. [30] assessed, in a population of 58 pediatric subjects, the agreement between the AHI from PSG and the 3% oxygen desaturation index (ODI3) from the $SpO_2$ signal simultaneously recorded with a portable monitoring device (SnoreSat, SagaTech Electronics, Calgary, AB, Canada). Similarly, Tsai et al. [32] evaluated the yield of several cut-offs for the 4% ODI (ODI4) to establish pediatric OSA severity in a database of 146 pediatric PSGs. Recently, Ma et al. [31] studied the clinical applicability of a pulse oximetry watch (POW) for pediatric OSA diagnosis. To this effect, they measured the concordance between the AHI from PSG and the ODI4 derived from the $SpO_2$ signal simultaneously recorded with the POW in 32 children.

Likewise, the number and depth of clusters of desaturations in the $SpO_2$ recordings have been quantified by means of a visual inspection in order to develop OSA screening protocols [20, 28, 33, 34]. A cluster of desaturations was defined by Brouillette et al. [28] as 5 or more oxygen desaturations of at least 4% occurring in a 10–30 minute window. Based on the number of clusters of desaturations and the number of drops of the $SpO_2$ signal below 90%, Brouillette et al. [28] defined a positive, negative, or inconclusive score for pediatric OSA and compared it with the standard diagnosis from PSG in a dataset of 349 children. Similarly, Nixon et al. [20] developed a severity score for pediatric OSA, named McGill oximetry score (MOS), which is also based on the number of clusters of desaturations and the number of drops of the oximetry signal below a defined threshold. Furthermore, Velasco et al. [34] assessed the diagnostic ability of a positive OSA score defined as 2 or more clusters of desaturations, one of them with an oxygen drop below 90%, in a sample of 167 children with adenotonsillar hypertrophy. Finally, Van Eyck et al. [33] assessed the diagnostic ability of the methodologies proposed by Brouillette et al. [28] and Velasco et al. [34], as well as the ODI3, in a population of 130 obese patients.

These oximetric indices have also been combined with common symptoms [29] and clinical history [35] in order to enhance their diagnostic ability. Chang et al. [29] used common symptoms (witnessed apneas, mouth breathing, and restless sleep) and the ODI3 to evaluate both a LR classifier and a discriminative score to diagnose pediatric OSA in a sample of 141 children. Conversely, Villa et al. [35] combined the MOS and a sleep clinical record that includes physical examination and children's history to detect pediatric OSA in a database of 236 pediatric subjects.

Differing from these studies, the research conducted in the present doctoral thesis has included the use of automated signal processing algorithms to further characterize the oximetry recordings.

### *1.5.2 Automated Signal Processing Methods*

Recent works have focused on enhancing the diagnostic capability of the oximetry signal by means of the application of automated signal processing algorithms [26, 36–41]. These studies typically follow a three-stage feature-engineering methodology. First, a feature extraction stage was performed to characterize the changes in oximetry dynamics associated to apneic events by means of different analytical approaches. Then, feature selection methods were applied to obtain optimum subsets of relevant and non-redundant features. Finally, pattern recognition algorithms were trained with the optimum subsets of features to detect pediatric OSA and its severity.

Importantly, in the feature extraction stage, the majority of studies employed signal processing techniques that had already shown its usefulness to extract features from the $SpO_2$ signal in adult patients both in the time and frequency domains [23]. In the time domain, $SpO_2$ recordings were characterized using statistical moments, oximetric indices, and several non-linear measures: approximate entropy [42], sample entropy [43], Lempel-Ziv complexity [44], and central tendency measure [45]. Conversely, the power spectral density (PSD) [46] was used to characterize the oximetry signal in the frequency domain [26, 37, 39–41].

First, Garde et al. [26] developed and validated the Phone Oximeter, a portable monitoring device consisting of a pulse oximetry sensor connected to a mobile phone, as a diagnostic tool for childhood OSA. To this effect, they assessed a LDA model fed with statistical parameters, classical indices, nonlinear features, and PSD-derived features from 146 $SpO_2$ recordings of pediatric patients. Similarly, Álvarez et al. [37] evaluated at-home unsupervised oximetry in pediatric OSA using a LR model trained with statistical moments, PSD-derived parameters, nonlinear features, and classical indices from 50 patients that underwent RP. A thorough comparative analysis of statistical binary classifiers for the diagnosis of childhood OSA was performed by Crepso et al. [39]. Specifically, they assessed LDA, LR, and quadratic discriminant analysis (QDA) pattern recognition models trained with statistical moments, PSD variables, nonlinear features, and the ODI3 extracted from a database of 176 children [39]. The usefulness of automated processing of oximetric recordings as a screening tool for pediatric OSA was also examined in a multicenter international study developed by Hornero et al. [40], which involved 4191 pediatric subjects from 13 sleep centers. Particularly, Hornero et al. [40] assessed a MLP neural network trained to estimate the AHI with the ODI3 and the third-order moment of the PSD. This MLP model was further validated in 432 children along with a remote cloud system [41].

Nonetheless, these studies used the same methods employed in adult OSA patients, but reaching a lower performance [26, 37, 39–41]. In this respect, Hornero et al. [40] showed a significant redundancy between the information extracted by conventional feature extraction methods and the ODI3. This highlights the need for novel signal processing methods able to provide specific features for the particularities of pediatric OSA.

In order to address this issue, two recent studies conducted by Crespo et al. [38] and Álvarez et al. [36] applied two novel nonlinear analysis methods in the context of pediatric OSA, multiscale entropy [47] and symbolic dynamics [48], respectively. Crespo et al. [38] investigated the ability of multiscale entropy (MSE) to further characterize the dynamics of unattended oximetry using the database employed in Álvarez et al. [37]. For this purpose, a LR model was trained to automatically detect childhood OSA with conventional oximetric indices and nonlinear MSE-derived parameters [38]. Additionally, Álvarez et al. [36] evaluated the usefulness of symbolic dynamics to increase the diagnostic capability of portable oximetry recordings from the Phone Oximeter. Specifically, a LR model was designed using conventional oximetric indices, anthropometric variable, statistical parameters, and nonlinear features from symbolic dynamics [36]. In this doctoral thesis, we have assessed the usefulness of detrended fluctuation analysis (DFA), a time domain nonlinear analysis method [49], and wavelet [50] and bispectral analysis [51], two frequency domain techniques, to provide complimentary features that help to further characterize apneic events from pediatric OSA [52–54].

As aforementioned, there has been a breakthrough in the last years in the data science field thanks to the emergence of deep learning approaches [6]. Due to its multilayer architecture with multiple levels of representation, deep-learning methods are suitable to learn very complex patterns from the raw data, which has led them to outperform conventional approaches in many fields [6]. In this respect, previous studies have applied deep-learning methods to analyze polysomnographical signals in adult OSA patients [7, 55]. These works have focused on the detection of sleep stages [7], apneic events [55], and/or direct AHI estimation [55]. To our knowledge, Vaquerizo-Villat et al. [56], the last article of the present doctoral thesis, is the first study that applies deep learning techniques in the context of childhood OSA. In Vaquerizo-Villat et al. [56], a deep-learning architecture based on convolutional neural networks (CNN) was fed with raw $SpO_2$ data [56], which has outperformed conventional approaches in the framework of pediatric OSA.

In this chapter, we have introduced the topic of this doctoral thesis. In the next Chapter (see Sect. 2), the hypotheses, objectives, and thematic consistency of this research will be stated.

## References

1. Bronzino JD (2000) Biomedical engineering handbook 2, vol. 2. Springer Science & Business Media
2. Sörnmo L, Laguna P (2005) Bioelectrical signal processing in cardiac and neurological applications. Academic Press
3. Najarian K, Splinter R (2012) Biomedical signal and image processing. Taylor & Francis
4. Krishnan S, Athavale Y (2018) Trends in biomedical signal feature extraction. Biomed Sig Process Control 43:41–63
5. Rangayyan RM (2015) Biomed Signal Anal. Wiley
6. LeCun Y, Bengio Y, Hinton G (2015) Deep learning. Nature 521(7553):436–444
7. Faust O, Razaghi H, Barika R, Ciaccio EJ, Acharya UR (2019) A review of automated sleep stage scoring based on physiological signals for the new millennia. Comput Meth Programs Biomed 176:81–91
8. Jahmunah V, Oh SL, Wei JKE, Ciaccio EJ, Chua K, San TR, Acharya UR (2019) Computer-aided diagnosis of congestive heart failure using ecg signals-a review. Phys Med 62:95–104
9. Roy Y, Banville H, Albuquerque I, Gramfort A, Falk TH, Faubert J (2019) Deep learning-based electroencephalography analysis: a systematic review. J Neural Eng 16(5):051001
10. Capdevila OS, Kheirandish-Gozal L, Dayyat E, Gozal D (2008) Pediatric obstructive sleep apnea: complications, management, and long-term outcomes. Proc Am Thorac Soc 5(2):274–282
11. Marcus CL, Brooks LJ, Ward SD, Draper KA, Gozal D, Halbower AC, Jones J, Lehmann C, Schechter MS, Sheldon S, Shiffman RN, Spruyt K (2012) Diagnosis and management of childhood obstructive sleep apnea syndrome. Pediatrics 130(3):e714–e755
12. Loughlin G, Brouillette R, Brooke L, Carroll J, Chipps B, England S, Ferber P, Ferraro N, Gaultier C, Givan D, Haddad GG (1996) Standards and indications for cardiopulmonary sleep studies in children. Am J Respir Crit Care Med 153(2):866–878
13. Hunter SJ, Gozal D, Smith DL, Philby MF, Kaylegian J, Kheirandish-Gozal L (2016) Effect of sleep-disordered breathing severity on cognitive performance measures in a large community cohort of young school-aged children. Am J Respir Crit Care Med 194(6):739–747
14. Alonso-Álvarez ML, Canet T, Cubell-Alarco M, Estivill E, Fernández E, Gozal D, Jurado-Luque MJ, Lluch-Roselló MA, Martínez-Pérez F, Merino-Andren M, Pin-Arboledas G, Roure N, Sanmartí FX, Sans-Capdevila O, Segarra-Isern J (2011) Documento de consenso del síndrome de apneas-hipopneas durante el sueño en niños. Arch Bronconeumol 47(Supl 5):2–18
15. Kheirandish-Gozal L (2010) What is abnormal in pediatric sleep? Respir Care 55(10):1366–1376
16. Tan H-L, Alonso Alvarez ML, Tsaoussoglou M, Weber S, Kaditis AG (2017) When and why to treat the child who snores? Pediatr Pulmonol 52(3):399–412
17. Tan H-L, Gozal D, Ramirez HM, Bandla HPR, Kheirandish-Gozal L (2014) Overnight polysomnography versus respiratory polygraphy in the diagnosis of pediatric obstructive sleep apnea. Sleep 37(2):255–260
18. Tan HL, Kheirandish-Gozal L, Gozal D (2015) Pediatric home sleep apnea testing slowly getting there! Chest 148(6):1382–1395
19. Katz ES, Mitchell RB, Ambrosio CMD (2012) Obstructive sleep apnea in infants. Am J Respir Crit care Med 185(8):805–816
20. Nixon GM, Kermack AS, Davis GM, Manoukian JJ, Brown A, Brouillette RT (2004) Planning adenotonsillectomy in children with obstructive sleep apnea: the role of overnight oximetry. Pediatrics 113(1):e19–e25
21. Kaditis AG, Alvarez MLA, Boudewyns A, Alexopoulos EI, Ersu R, Joosten K, Larramona H, Miano S, Narang I, Trang H, Tsaoussoglou M, Vandenbussche N, Villa MP, Waardenburg DV, Weber S, Verhulst S (2016) Obstructive sleep disordered breathing in 2- to 18-year-old children: diagnosis and management. Eur Respir J 47(1):69–94

22. Standards of Practice Committee of the American Sleep Disorders Association (1994) Practice parameters for the use of portable recording in the assessment of obstructive sleep apnea. Sleep 17(4):372–377
23. del Campo F, Crespo A, Cerezo-Hernández A, Gutiérrez-Tobal GC, Hornero R, Álvarez D (2018) Oximetry use in obstructive sleep apnea. Expert Rev Respir Med 12(8):665–681
24. Netzer N, Eliasson AH, Netzer C, Kristo DA (2001) Overnight pulse oximetry for sleep-disordered breathing in adults: a review. Chest 120(2):625–633
25. Chan ED, Chan MM, Chan MM (2013) Pulse oximetry: understanding its basic principles facilitates appreciation of its limitations. Respir Med 107(6):789–799
26. Garde A, Dehkordi P, Karlen W, Wensley D, Ansermino JM, Dumont GA (2014) Development of a screening tool for sleep disordered breathing in children using the phone oximeter. PloS one 9(11):e112959
27. Berry RB, Budhiraja R, Gottlieb DJ, Gozal D, Iber C, Kapur VK, Marcus CL, Mehra R, Parthasarathy S, Quan SF, Redline S, Strohl KP, Davidson Ward SL, Tangredi MM (2012) Rules for scoring respiratory events in sleep: update of the 2007 AASM manual for the scoring of sleep and associated events: deliberations of the sleep apnea definitions task force of the American Academy of Sleep Medicine. J Clin Sleep Med 8(5):597
28. Brouillette RT, Morielli A, Leimanis A, Waters KA, Luciano R, Ducharme FM (2000) Nocturnal pulse oximetry as an abbreviated testing modality for pediatric obstructive sleep apnea. Pediatrics 105(2):405–412
29. Chang L, Wu J, Cao L (2013) Combination of symptoms and oxygen desaturation index in predicting childhood obstructive sleep apnea. Int J Pediatr Otorhinolaryngol 77(3):365–371
30. Kirk VG, Bohn SG, Flemons WW, Remmers JE (2003) Comparison of home oximetry monitoring with laboratory polysomnography in children. Chest 124(5):1702–1708
31. Ma J-R, Huang J-J, Chen Q, Wu H-T, Xiao K-L, Zhang Y-T (2018) Value of pulse oximetry watch for diagnosing pediatric obstructive sleep apnea/hypopnea syndrome. Acta oto-laryngol 138(2):175–179
32. Tsai C-M, Kang C-H, Su M-C, Lin H-C, Huang E-Y, Chen C-C, Hung J-C, Niu C-K, Liao D-L, Yu H-R (2013) Usefulness of desaturation index for the assessment of obstructive sleep apnea syndrome in children. Int J Pediatr Otorhinolaryngol 77(8):1286–1290
33. Van Eyck A, Lambrechts C, Vanheeswijck L (2015) The role of nocturnal pulse oximetry in the screening for obstructive sleep apnea in obese children and adolescents. Sleep Med 16(11):1409–1412
34. Velasco CT, Suárez MD, Figueroa JM, Turienzo MD, Len FL, Mansilla E (2013) Pulse oximetry recording in children with adenotonsillar hypertrophy?: usefulness in the diagnosis of obstructive sleep apnea syndrome. Arch argentinos de Pediatr 111(3):196–201
35. Villa MP, Pietropaoli N, Supino MC, Vitelli O, Rabasco J, Evangelisti M, Del Pozzo M, Kaditis AG (2015) Diagnosis of pediatric obstructive sleep apnea syndrome in settings with limited resources. JAMA Otolaryngol-Head Neck Surg 141(11):990–996
36. Álvarez D, Crespo A, Vaquerizo-Villar F, Gutiérrez-Tobal GC, Cerezo-Hernández A, Barroso-García V, Ansermino JM, Dumont GA, Hornero R, Del Campo F, Garde A (2018) Symbolic dynamics to enhance diagnostic ability of portable oximetry from the phone oximeter in the detection of paediatric sleep apnoea. Phys Measur 39(10):104002
37. Álvarez D, Alonso-Álvarez ML, Gutiérrez-Tobal GC, Crespo A, Kheirandish-Gozal L, Gozal D, Terán-Santos J, Campo FD (2017) Automated screening of children with obstructive sleep apnea using nocturnal oximetry: an alternative to respiratory polygraphy in unattended settings. J Clin Sleep Med 13(5):7–11
38. Crespo A, Álvarez D, Gutiérrez-Tobal GC, Vaquerizo-Villar F, Barroso-García V, Alonso-Álvarez ML, Terán-Santos J, Hornero R, del Campo F (2017) Multiscale entropy analysis of unattended oximetric recordings to assist in the screening of paediatric sleep apnoea at home. Entropy 19(6):284
39. Crespo A, Álvarez D, Kheirandish-Gozal L, Gutiérrez-Tobal GC, Cerezo-Hernández A, Gozal D, Hornero R, Del Campo F (2018) Assessment of oximetry-based statistical classifiers as simplified screening tools in the management of childhood obstructive sleep apnea. Sleep Breathing 22(4):1063–1073

40. Hornero R, Kheirandish-Gozal L, Gutiérrez-Tobal GC, Philby MF, Alonso-Álvarez ML, Álvarez D, Dayyat EA, Xu Z, Huang Y-S, Tamae Kakazu M, Li AM, Van Eyck A, Brockmann PE, Ehsan Z, Simakajornboon N, Kaditis AG, Vaquerizo-Villar F, Crespo Sedano A, Sans Capdevila O, von Lukowicz M, Terán-Santos J, Del Campo F, Poets CF, Ferreira R, Bertran K, Zhang Y, Schuen J, Verhulst S, Gozal D (2017) Nocturnal oximetry-based evaluation of habitually snoring children. Am J Respir Crit Care Med 196(12):1591–1598
41. Xu Z, Gutiérrez-Tobal GC, Wu Y, Kheirandish-Gozal L, Ni X, Hornero R, Gozal D (2019) Cloud algorithm-driven oximetry-based diagnosis of obstructive sleep apnoea in symptomatic habitually snoring children. Eur Respir J 53(2)
42. Pincus SM (1991) Approximate entropy as a measure of system complexity. Proc National Acad Sci 88(6):2297–2301
43. Richman JS, Moorman JR (2000) Physiological time-series analysis using approximate entropy and sample entropy. Am J Physiol-Heart Circulatory Physiol 278(6):H2039–H2049
44. Lempel A, Ziv J (1976) On the complexity of finite sequences. IEEE Trans Inf Theory 22(1):75–81
45. Cohen ME, Hudson DL (2000) New chaotic methods for biomedical signal analysis. In: IEEE EMBS international conference on information technology applications in biomedicine. pp 123–128
46. Welch PD (1967) The use of fast fourier transform fot he estimation of power spectra: a method based on time averaging over short, modified periodograms. IEEE Trans Audio Electroacoust 15(2):70–73
47. Costa M, Goldberger AL, Peng C (2005) Multiscale entropy analysis of biological signals. Phys Rev E 71(2):1–18
48. Daw CS, Finney CEA, Tracy ER (2003) A review of symbolic analysis of experimental data. Rev Sci instrum 74(2):915–930
49. Peng C-K, Buldyrev SV, Havlin S, Simons M, Stanley HE, Goldberger AL (1994) Mosaic organization of DNA nucleotides. Phys Rev E 49(2):1685
50. Rioul O, Vetterli M (1991) Wavelets and signal processing. IEEE Signal Process Mag 8(4):14–38
51. Chua K, Chandran V, Acharya UR, Min C (2010) Application of higher order statistics/spectra in biomedical signals a review. Med Eng Phys 32(7):679–689
52. Vaquerizo-Villar F, Álvarez D, Kheirandish-Gozal L, Gutiérrez-Tobal GC, Barroso-García V, Crespo A, del Campo F, Gozal D, Hornero R (2018) Utility of bispectrum in the screening of pediatric sleep apnea-hypopnea syndrome using oximetry recordings. Comput Meth Programs Biomed 156:141–149
53. Vaquerizo-Villar F, Álvarez D, Kheirandish-Gozal L, Gutiérrez-Tobal GC, Barroso-García V, Crespo A, del Campo F, Gozal D, Hornero R (2018) Wavelet analysis of oximetry recordings to assist in the automated detection of moderate-to-severe pediatric sleep apnea-hypopnea syndrome. PloS one 13(12):e0208502
54. Vaquerizo-Villar F, Álvarez D, Kheirandish-Gozal L, Gutiérrez-Tobal GC, Barroso-García V, Crespo A, Del Campo F, Gozal D, Hornero R (2018) Detrended fluctuation analysis of the oximetry signal to assist in paediatric sleep apnoea-hypopnoea syndrome diagnosis. Physiol Meas 39(11):114006
55. Mostafa SS, Mendonça F, Ravelo-García G (2019) A systematic review of detecting sleep apnea using deep learning. Sensors 19(22):4934
56. Vaquerizo-Villar F, Alvarez D, Kheirandish-Gozal L, Gutierrez-Tobal GC, Barroso-Garcia V, Santamaria-Vazquez E, Del Campo F, Gozal D, Hornero R (2021) A convolutional neural network architecture to enhance oximetry ability to diagnose pediatric obstructive sleep apnea. IEEE J Biomed Health Inform 25(8), 2906–2916

# Chapter 2
# Hypotheses and Objectives

Pediatric OSA is a high prevalent disease (1–5%) [1]. It is associated with many negative effects on the overall health and life quality of the affected children when it is untreated, including cardiometabolic malfunctioning and neurobehavioral abnormalities [2]. Overnight PSG is the gold standard for pediatric OSA diagnosis [1]. Despite its effectiveness, PSG is costly, complex, highly intrusive, and lacks availability [3, 4]. This has prompted the search for simplified screening tests [1, 5]. One of these alternative tests is overnight oximetry, which measures the $SpO_2$ signal with a pulse-oximeter, typically placed onto the end of a finger, thus being especially suitable for children [6]. A wide range of investigations have shown the utility of the automated analysis of the $SpO_2$ signal as a clinically beneficial tool for the screening of OSA in adult patients [6]. Nonetheless, state-of-the-art studies focused on the automated analysis of the oximetry signal in the context of pediatric OSA followed a similar signal-processing methodology [6], but achieving a inferior diagnostic performance than in the case of adult patients [6].

Accordingly, the present doctoral thesis focuses on the use of feature-engineering and deep-learning methodologies, intended to increase the diagnosis ability of the $SpO_2$ signal from overnight oximetry. Therefore, Sect. 2.1 describes the different hypothesis that have been formulated throughout the present doctoral thesis, as well as the global hypothesis that raised from these statements. Similarly, the main objective of this thesis is stated in Sect. 2.2, as well as the specific objectives that have been accomplished to achieve it. Finally, thematic consistency of the four scientific articles that have been accepted or published in journals indexed in the Journal Citation Reports (JCR) from the Web of Science™ and are included in this doctoral thesis is explained in Sect. 2.3.

 
F. Vaquerizo Villar, *Automated Analysis of the Oximetry Signal to Simplify the Diagnosis of Pediatric Sleep Apnea*, Springer Theses,
https://doi.org/10.1007/978-3-031-32832-9_2

## 2.1 Hypotheses

Simplification of OSA diagnosis has become a main research topic in past years [1, 5, 7]. As previously explained, the $SpO_2$ signal allows to detect oxygen desaturations associated to apneic events [8], which has led to its use in the screening of pediatric OSA [6]. Thus, at the beginning of this doctoral thesis, the following hypothesis was formulated: *the oximetry signal on its own may contain enough information for the screening of pediatric OSA*. Nonetheless, this statement does not completely describe the starting point of the different investigations conducted in this study.

As stated in Sect. 1.5.2, a recent study reported a high redundancy in the conventional features extracted from the oximetry signal [9], suggesting the need to apply novel signal processing algorithms. Thereby, it has been hypothesized that *novel feature extraction methods could further characterize OSA-related changes in the oximetry signal*. Similarly, we wonder whether *the features extracted by these methods provide complimentary information to improve the diagnostic capability of the oximetry signal in the context of pediatric OSA*. Feature selection and pattern recognition algorithms have been used for this purpose.

In spite of the usefulness of conventional feature-engineering approaches, they are limited to learn all the relevant information from the data because of: (i) the domain expert that designs the feature extractor decides which features are relevant a priori; and (ii) their reduced level of abstraction. Owing to their great capability to automatically learn very complex features from raw data, it has been hypothesized that *deep-learning algorithms could extract all the OSA-related information from the* $SpO_2$ *signal, thus enhancing its diagnostic ability*.

These statements are the main hypotheses that form the core of the current doctoral thesis, which can be joint into the following global hypothesis:

> The application of novel feature extraction and deep-learning algorithms allows to capture hidden patterns of desaturations linked to apneic events, enhancing the diagnostic ability of the single-channel oximetry in the context of pediatric OSA.

## 2.2 Objectives

The main goal of the present doctoral thesis is to design, develop, and evaluate novel clinical decision-support models in the context of pediatric OSA based on the automated analysis of the oximetry signal. To reach this general goal, the following specific objectives arise:

I. To further characterize changes in the oximetry signal caused by apneic events linked to pediatric OSA both in the time and frequency domains.
II. To identify novel features from the oximetry signal able to provide relevant and complimentary information to conventional oximetry variables.

III. To design and optimize high-performance pattern recognition models aimed at the automated detection of pediatric OSA and its severity using optimum subsets of features from the $SpO_2$ signal.
IV. To explore and develop novel deep-learning based architectures capable to automatically learn all the OSA-related information from raw oximetry data.

## 2.3 Thesis Organization

The automated analysis of the oximetry signal has become the primary alternative to PSG in the screening of pediatric OSA. In this context, the present doctoral thesis focuses on applying novel signal processing algorithms in order to enhance the diagnostic ability of the oximetry signal in the framework of pediatric OSA. All the papers included in this thesis share this common thread. Figure 2.1 shows the thematic consistency and the main contributions of the papers included in this thesis.

As mentioned in Sect. 1.5.2, earlier studies focused on the automated diagnosis of childhood OSA from the oximetry signal followed a similar feature-engineering methodology to that used in adult patients, but achieving an inferior diagnostic performance [6]. Specifically, a previous study using a very large multicenter database of 4191 PSG sleep studies reported a high redundancy in the information extracted from the oximetry recordings [9], which suggests the need to seek novel signal processing algorithms that provide additional information from the $SpO_2$ signal for the particularities of childhood OSA (see Sect. 1.5.2). In this respect, the first three papers in chronological order were focused on the application of novel feature extraction algorithms to provide additional information from the oximetry signal [10–12]. Given that the information extracted by conventional spectral analysis techniques is highly redundant, the first two papers were aimed at assessing the usefulness of two frequency domain techniques, bispectrum [10] and wavelet analysis [11], to provide discriminative frequency domain features from the oximetry signal. In the third paper [12], we investigated if DFA, a nonlinear analysis method in the time domain, can extract complimentary information from the oximetry signal linked to apneic events. As we will see, these feature extraction methods, novel in the context of pediatric OSA, have been found to provide complimentary information to improve the diagnosis ability of the $SpO_2$ signal.

In contrast to the first three papers, which followed a feature-engineering methodology, the last paper included in this thesis employed a deep-learning methodology to analyze the oximetry signal. Deep-learning approaches have emerged in the last years as a suitable tool to learn complex features from raw data using architectures with multiple layers of representation [13]. These algorithms have outperformed traditional feature-engineering approaches in many fields, including image recognition, natural language processing, and time series analysis [13]. Accordingly, in the last paper we investigated the ability of CNNs, the most widely-used deep-learning technique, to automatically extract all the relevant information from the oximetry signal [14]. The proposed CNN architecture was validated in a multicenter database

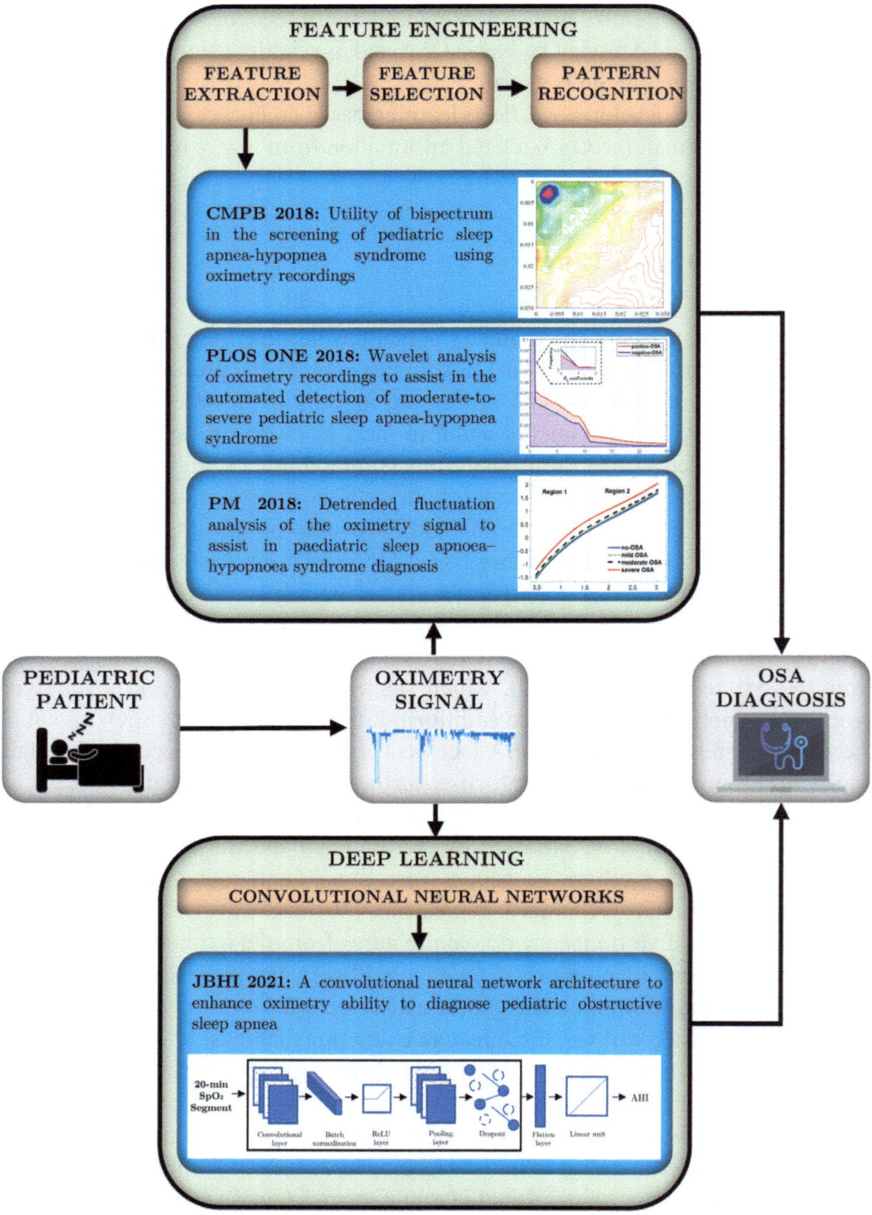


**Fig. 2.1** Main contributions of the papers included in the doctoral thesis, arranged along the automated signal processing methodologies developed. CMPB: Computer Methods and Programs in Biomedicine, IEEE JBHI: IEEE Journal of Biomedical and Health Informatics, PM: Physiological Measurement

of 3196 $SpO_2$ recordings, showing a high diagnostic ability, which outperformed conventional feature-engineering approaches.

Titles, authors, and abstracts of each of the papers of the present doctoral thesis, as well as the indexed journals in which they were accepted and published are shown below:

**Utility of bispectrum in the screening of pediatric sleep apnea-hypopnea syndrome using oximetry recordings [10].**

**Fernando Vaquerizo-Villar**, Daniel Álvarez, Leila Kheirandish-Gozal, Gonzalo C. Gutiérrez-Tobal, Verónica Barroso-García, Andrea Crespo, Félix del Campo, David Gozal, and Roberto Hornero. *Computer Methods and Programs in Biomedicine*, vol. 156, p. 141–149, 2018. Impact factor in 2018: 3.424, Q1 in "COMPUTER SCIENCE, THEORY & METHODS" (JCR-WOS).

**Wavelet analysis of oximetry recordings to assist in the automated detection of moderate-to-severe pediatric sleep apnea-hypopnea syndrome [11].**

**Fernando Vaquerizo-Villar**, Daniel Álvarez, Leila Kheirandish-Gozal, Gonzalo C. Gutiérrez-Tobal, Verónica Barroso-García, Andrea Crespo, Félix del Campo, David Gozal, and Roberto Hornero. *PLOS One*, vol. 13 (12), p. e0208502, 2018. Impact factor in 2018: 2.776, Q2 in "MULTIDISCIPLINARY SCIENCES" (JCR-WOS).

**Detrended fluctuation analysis of the oximetry signal to assist in paediatric sleep apnoea hypopnoea syndrome diagnosis [12].**

**Fernando Vaquerizo-Villar**, Daniel Álvarez, Leila Kheirandish-Gozal, Gonzalo C. Gutiérrez-Tobal, Verónica Barroso-García, Andrea Crespo, Félix del Campo, David Gozal, and Roberto Hornero. *Physiological Measurement*, vol. 39 (11), p. 114006, 2018. Impact factor in 2018: 2.246, Q3 in "ENGINEERING, BIOMEDICAL" (JCR-WOS).

**A convolutional neural network architecture to enhance oximetry ability to diagnose pediatric obstructive sleep apnea [14].**

Fernando Vaquerizo-Villar, Daniel Álvarez, Leila Kheirandish-Gozal, Gonzalo C. Gutiérrez-Tobal, Verónica Barroso-García, Eduardo Santamaría-Vázquez, Félix del Campo, David Gozal, and Roberto Hornero. *IEEE Journal of Biomedical and Health Informatics*, vol. 25 (8), p. 2906–2916, 2021. Impact factor in 2021: 7.021, D1 in "MATHEMATICAL & COMPUTATIONAL BIOLOGY" (JCR-WOS).

## References


1. Marcus CL, Brooks LJ, Ward SD, Draper KA, Gozal D, Halbower AC, Jones J, Lehmann C, Schechter MS, Sheldon S, Shiffman RN, Spruyt K (2012) Diagnosis and management of childhood obstructive sleep apnea syndrome. Pediatrics 130(3):e714–e755
2. Capdevila OS, Kheirandish-Gozal L, Dayyat E, Gozal D (2008) Pediatric obstructive sleep apnea: complications, management, and long-term outcomes. Proc Am Thorac Soc 5(2):274–282
3. Katz ES, Mitchell RB, Ambrosio CMD (2012) Obstructive sleep apnea in infants. Am J Respir Crit Care Med 185(8):805–816
4. Tan HL, Kheirandish-Gozal L, Gozal D (2015) Pediatric home sleep apnea testing slowly getting there! Chest 148(6):1382–1395
5. Kaditis AG, Alvarez MLA, Boudewyns A, Alexopoulos EI, Ersu R, Joosten K, Larramona H, Miano S, Narang I, Trang H, Tsaoussoglou M, Vandenbussche N, Villa MP, Waardenburg DV, Weber S, Verhulst S (2016) Obstructive sleep disordered breathing in 2- to 18-year-old children: diagnosis and management. Eur Respir J 47(1):69–94
6. del Campo F, Crespo A, Cerezo-Hernández A, Gutiérrez-Tobal GC, Hornero R, Álvarez D (2018) Oximetry use in obstructive sleep apnea. Expert Rev Respir Med 12(8):665–681
7. Bertoni D, Isaiah A (2019) Towards patient-centered diagnosis of pediatric obstructive sleep apnea a review of biomedical engineering strategies. Expert Rev Med Devices 16(7):617–629
8. Berry RB, Budhiraja R, Gottlieb DJ, Gozal D, Iber C, Kapur VK, Marcus CL, Mehra R, Parthasarathy S, Quan SF, Redline S, Strohl KP, Davidson Ward SL, Tangredi MM (2012) Rules for scoring respiratory events in sleep: update of the 2007 AASM manual for the scoring of sleep and associated events: deliberations of the sleep apnea definitions task force of the American Academy of Sleep Medicine. J Clin Sleep Med 8(5):597
9. Hornero R, Kheirandish-Gozal L, Gutiérrez-Tobal GC, Philby MF, Alonso-Álvarez ML, Álvarez D, Dayyat EA, Xu Z, Huang Y-S, Tamae Kakazu M, Li AM, Van Eyck A, Brockmann PE, Ehsan Z, Simakajornboon N, Kaditis AG, Vaquerizo-Villar F, Crespo Sedano A, Sans Capdevila O, von Lukowicz M, Terán-Santos J, Del Campo F, Poets CF, Ferreira R, Bertran K, Zhang Y, Schuen J, Verhulst S, Gozal D (2017) Nocturnal oximetry-based evaluation of habitually snoring children. Am J Respir Crit Care Med 196(12):1591–1598
10. Vaquerizo-Villar F, Álvarez D, Kheirandish-Gozal L, Gutiérrez-Tobal GC, Barroso-García V, Crespo A, del Campo F, Gozal D, Hornero R (2018) Utility of bispectrum in the screening of pediatric sleep apnea-hypopnea syndrome using oximetry recordings. Comput Methods Programs Biomed 156:141–149
11. Vaquerizo-Villar F, Álvarez D, Kheirandish-Gozal L, Gutiérrez-Tobal GC, Barroso-García V, Crespo A, del Campo F, Gozal D, Hornero R (2018) Wavelet analysis of oximetry recordings to assist in the automated detection of moderate-to-severe pediatric sleep apnea-hypopnea syndrome. PloS One 13(12):e0208502
12. Vaquerizo-Villar F, Álvarez D, Kheirandish-Gozal L, Gutiérrez-Tobal GC, Barroso-García V, Crespo A, Del Campo F, Gozal D, Hornero R (2018) Detrended fluctuation analysis of the oximetry signal to assist in paediatric sleep apnoea-hypopnoea syndrome diagnosis. Physiol Meas 39(11):114006
13. LeCun Y, Bengio Y, Hinton G (2015) Deep learning. Nature 521(7553):436–444
14. Vaquerizo-Villar F, Alvarez D, Kheirandish-Gozal L, Gutierrez-Tobal GC, Barroso-Garcia V, Santamaria-Vazquez E, Del Campo F, Gozal D, Hornero R (2021) A convolutional neural network architecture to enhance oximetry ability to diagnose pediatric obstructive sleep apnea. IEEE J Biomed Health Inform 25(8), 2906–2916

# Chapter 3
# Methods

This chapter first shows a brief summary the databases of pediatric subjects used (Sect. 3.1) and then describes the different stages of the general signal processing methodology that has been conducted through the thesis (see Fig. 3.1). The methodology starts with a signal pre-processing stage (Sect. 3.2), which adapts the oximetry data to the requirements of the different signal processing algorithms. Following the feature-engineering branch (Sect. 3.3), Sects. 3.3.1 and 3.3.2 describe the signal processing methods applied to extract features from the $SpO_2$ recordings in the time and frequency domains and to select optimum subsets of optimum OSA-related features, respectively. Afterward, Sect. 3.3.3 is devoted to describe the pattern recognition algorithms employed to establish pediatric OSA and its severity. As the last stage of the feature-engineering methodology, pattern recognition algorithms are fed with the optimum features obtained in the feature selection stage. In the deep-learning branch (Sect. 3.4), CNNs are trained using raw oximetry signal to estimate the AHI and thereby the severity of pediatric OSA. Finally, Sect. 3.5 describes the statistical analysis techniques employed in this research.

## 3.1 Subjects and Signals Under Study

During this research, three different databases of pediatric subjects were analyzed: (i) the Childhood Adenotonsillectomy Trial (CHAT) database, (ii) the University of Chicago (UofC) database, and (iii) the Burgos University Hospital (BUH) database. All of them contained $SpO_2$ recordings of children ranging from 0 to 18 years of age. These pediatric subjects were referred to nocturnal PSG showing clinical suspicion from OSA due to one or several of the following criteria: snoring, apneas, arousals, excessive daytime sleepiness, restless sleep, hyperactivity, tonsillar hypertrophy, increase in neck circumference, developmental disorder depression and low self-esteem, enuresis, obesity, attention deficit, behavioral problems, and cephaleas.



F. Vaquerizo Villar, *Automated Analysis of the Oximetry Signal to Simplify the Diagnosis of Pediatric Sleep Apnea*, Springer Theses,
https://doi.org/10.1007/978-3-031-32832-9_3

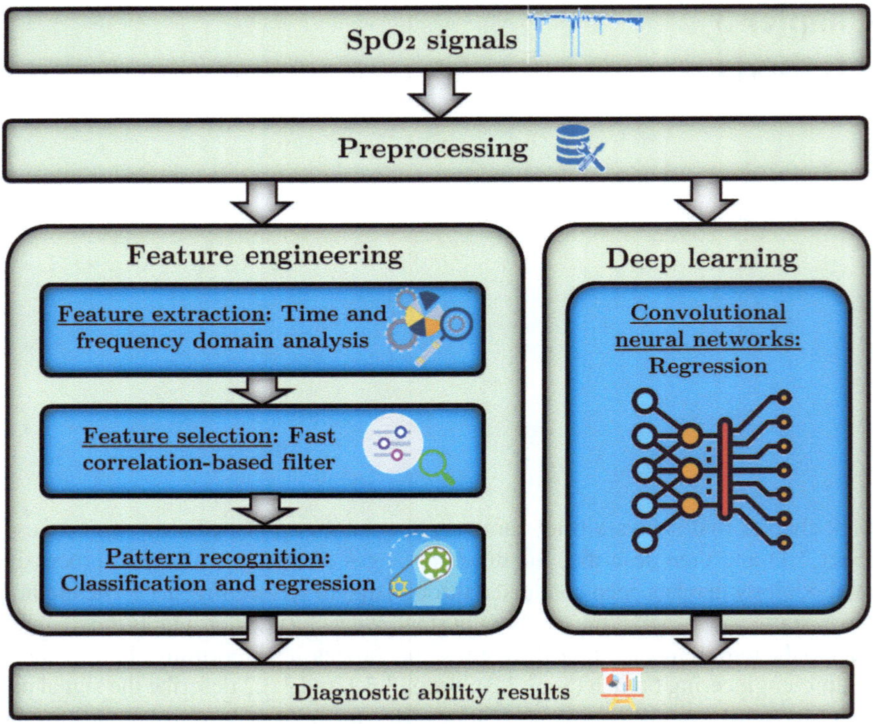


**Fig. 3.1** Scheme of the general signal processing methodology conducted in the study

The CHAT dataset, a public multicenter dataset, was integrated by 1638 sleep studies. The UofC dataset consisted of 980 children and the BUH dataset was composed of 578 pediatric subjects.

$SpO_2$ recordings were acquired from pediatric subjects during their standard diagnostic PSG test using sampling frequencies ranging from 1 to 512 Hz. Following the rules of the AASM [1, 2], all the sleep recordings were manually inspected by trained staff to quantify sleep and annotate apneas and hypopneas. These annotations were used to obtain the AHI, which is used to establish a diagnosis [3]. Common AHI cutoffs used to establish pediatric OSA severity are 1, 5, and 10 e/h [4–6]. In this respect, AHI = 5 e/h is commonly employed as a cutoff to recommend surgical treatment [6], as children with an AHI $\geq$ 5 e/h are at a higher risk of developing comorbidities [5, 7, 8]. Accordingly, under a binary classification approach, pediatric subjects were distinguished into negative OSA (AHI < 5 e/h) and positive OSA (AHI $\geq$ 5 e/h). In addition, children can also be classified into four pediatric OSA severity degrees: no OSA (AHI < 1 e/h), mild OSA (1 $\leq$ AHI < 5 e/h), moderate OSA (5 $\leq$ AHI < 10 e/h), and severe OSA (AHI $\geq$ 10 e/h). Details of each database, such as number of subjects, sex (male percentage), age, body mass index (BMI), and number of patients of each OSA severity group are provided in Tables 3.1, 3.2, 3.3, and 3.4.

**Table 3.1** Clinical and sociodemographic data of the CHAT database

| | All | No OSA | Mild OSA | Moderate OSA | Severe OSA |
|---|---|---|---|---|---|
| $SpO_2$ recordings (%) | 1638 (100%) | 637 (38.9%) | 609 (37.2%) | 205 (12.5%) | 187 (11.4%) |
| Age (years) | 7 [6, 8] | 7 [6, 8] | 7 [6, 8] | 7 [6, 8] | 7 [6, 8] |
| Males (%) | 602 (47.4%) | 297 (46.6%) | 287 (47.1%) | 101 (49.3%) | 92 (49.2%) |
| BMI (kg/m$^2$) | 17.3 [15.8, 21.7] | 17.0 [15.5, 19.6] | 17.4 [15.6, 21.7] | 18.6 [15.4, 23.3] | 18.9 [16.0, 24.3] |
| AHI (e/h) | 1.6 [0.6, 4.7] | 0.4 [0.2, 0.7] | 2.2 [1.5, 3.2] | 7.1 [5.9, 8.4] | 17.9 [12.8, 26.9] |

Data are presented as median [interquartile range], *n*, or *n* (%). OSA Obstructive sleep apnea

### *3.1.1 Childhood Adenotonsillectomy Trial (CHAT) Database*

CHAT database is composed of 1639 sleep studies from children ranging 5–10 years old with clinical suspicion of OSA. Sleep studies were obtained in 6 pediatric sleep centers of the United States of America (Children's Hospital of Pennsylvania, Philadelphia, PA; Cincinnati Children's Medical Center, Cincinnati, OH; Rainbow Babies and Children's Hospital, Cleveland, OH; Boston Children's Hospital, Boston, MA; Cardinal Glennon Children's Hospital, St. Louis, MI; Montefiore Medical Center, Bronx, NY) of which 1638 contained $SpO_2$ recording [9, 10]. This database is divided into three groups:

- Baseline, composed of 453 $SpO_2$ recordings from children that met the criteria defined in Marcus et al. [9] and Redline et al. [10] to be randomized to early adenotonsillectomy or a strategy of watchful waiting.
- Follow-up, composed of $SpO_2$ recordings from 406 pediatric subjects of the baseline group performed after a 7-month observation period.
- Nonrandomized, composed of 779 $SpO_2$ recordings from children who did not met the criteria defined in Marcus et al. [9] and Redline et al. [10] to be included in the follow-up study.

Overnight PSG was performed following a strict standardized procedure [10], which includes the acquisition of $SpO_2$ recordings with a Nonin 8000J or comparable sensor at sampling rates ranging from 1 to 512 Hz. The clinical trail identifier of the CHAT database is available in NCT00560859 and its full protocol is provided in the supplementary material of Marcus et al. [9]. This database was used in Vaquerizo-Villar et al. [11]. Table 3.1 shows the sociodemographic and clinical data from this database.

### *3.1.2 University of Chicago (UofC) Database*

UofC database is composed of 981 $SpO_2$ recordings from children aged 0–13 years of age. All children were referred to the pediatric sleep unit at the University of Chicago Medicine Comer Children's Hospital (Chicago, IL, USA) due to clinical

**Table 3.2** Clinical and sociodemographic data of the complete UofC database (981 subjects)

| | All | No OSA | Mild OSA | Moderate OSA | Severe OSA |
|---|---|---|---|---|---|
| $SpO_2$ recordings (%) | 981 (100%) | 175 (17.8%) | 401 (40.9%) | 176 (17.9%) | 229 (23.4%) |
| Age (years) | 6 [3, 9] | 7 [4, 10] | 6 [4, 9] | 5 [2, 8] | 4 [2, 8] |
| Males (%) | 602 (61.4%) | 109 (62.3%) | 247 (61.6%) | 107 (60.8%) | 139 (60.7%) |
| BMI ($kg/m^2$) | 17.9 [15.8, 21.9] | 17.7 [15.5, 20.9] | 17.7 [15.9, 21.2] | 18.6 [16.2, 24.0] | 18.3 [16.0, 23.2] |
| AHI (e/h) | 3.8 [1.5, 9.3] | 0.5 [0.1, 0.8] | 2.5 [1.7, 3.5] | 6.8 [5.8, 8.3] | 19.1 [13.9, 31.1] |

Data are presented as median [interquartile range], *n*, or *n* (%). OSA Obstructive sleep apnea

**Table 3.3** Clinical and sociodemographic data of the initial version of the UofC database (298 subjects)

| | All | Negative OSA | Moderate OSA | Severe OSA |
|---|---|---|---|---|
| $SpO_2$ recordings (%) | 298 (100%) | 164 (55.0%) | 56 (18.8%) | 78 (26.2%) |
| Age (years) | 6 [4, 9] | 7 [5, 10] | 5 [3, 8] | 6 [3, 9] |
| Males (%) | 166 (55.7%) | 91 (55.5%) | 32 (57.1%) | 43 (55.1%) |
| BMI ($kg/m^2$) | 18.4 [16.3, 23.0] | 18.2 [16.3, 22.3] | 18.1 [16.3, 22.6] | 19.1 [16.5, 25.7] |
| AHI (e/h) | 4.2 [1.8, 10.4] | 1.9 [1.0, 3.5] | 7.0 [5.9, 8.5] | 17.7 [11.7, 27.3] |

Data are presented as median [interquartile range], *n*, or *n* (%). OSA Obstructive sleep apnea

**Table 3.4** Clinical and sociodemographic data of the BUH database

| | All | No OSA | Mild OSA | Moderate OSA | Severe OSA |
|---|---|---|---|---|---|
| $SpO_2$ recordings (%) | 578 (100%) | 205 (35.5%) | 220 (38.1%) | 65 (11.3%) | 88 (15.2%) |
| Age (years) | 5 [4, 7] | 6 [4, 8] | 5 [3, 6] | 5 [3, 6] | 4 [3, 5] |
| Males (%) | 356 (61.6%) | 127 (62.0%) | 129 (58.7%) | 38 (58.5%) | 62 (70.5%) |
| BMI ($kg/m^2$) | 16.0 [14.6, 18.2] | 16.1 [14.5, 18.8] | 16.0 [14.7, 17.7] | 15.4 [14.6,18.1] | 16.1 [14.7,17.3] |
| AHI (e/h) | 1.8 [0.6, 5.3] | 0.4 [0.0, 0.6] | 2.1 [1.5, 3.4] | 6.9 [5.8, 8.1] | 24.3 [14.8, 34.9] |

Data are presented as median [interquartile range], *n*, or *n* (%). OSA Obstructive sleep apnea

suspicion of OSA. The legal guardians of all the children signed the informed consent and the Ethics Committee of the Comer Children's Hospital approved the protocols (#11-0268-AM017, #09-115-B-AM031, and #IRB14-1241). Overnight PSGs were performed using a digital polysomnography system (Polysmith; Nihon Kohden America Inc., CA, USA), which includes an internal pulse oximetry sensor. In this way, $SpO_2$ recordings were obtained from PSG at sampling rates 25, 200, 500 Hz.

This dataset was used in all the studies of this doctoral thesis [11–14]. In the first manuscript of this doctoral thesis [12], a subset of the whole UofC database composed of 298 $SpO_2$ recordings, those sampled 25 Hz, was analyzed. In Vaquerizo-Villar et al. [13, 14], the complete database, 981 subjects, was employed. Finally, one subject was removed from the complete dataset for the last manuscript of this thesis, as the total sleep time was less than 3 h after signal preprocessing, thus analyzing 980 $SpO_2$ recordings in Vaquerizo-Villar et al. [11]. Tables 3.2 and 3.3 show the clinical and sociodemographic data from the complete dataset (981 subjects) and the initial dataset of 298 $SpO_2$ recordings used in Vaquerizo-Villar et al. [12], respectively.

### *3.1.3 Burgos University Hospital (BUH) Database*

The BUH database included 578 children ranging 0–18 years of age who were referred to the pediatric sleep laboratory at Burgos University Hospital with high suspicion from OSA. All legal guardians of the pediatric subjects involved gave their informed consent and the Ethics Committee of the BUH approved the protocol. Children's sleep was monitored using the Deltamed Coherence 3NT Polysomnograph, version 3.0 system (Diagniscan S.A.U., Group Werfen, Paris, France), which includes a Nellcor Puritan Bennett, NPB-290 pulse oximeter. In this way, $SpO_2$ recordings were obtained during nocturnal PSG at a sampling rate 200 Hz. This database was used in Vaquerizo-Villar et al. [11]. Table 3.4 shows the sociodemographic and clinical data from this sample.

## 3.2 Pre-processing

As seen in Sect. 3.1, $SpO_2$ recordings were acquired during PSG using different pulse oximetry sensors and recording systems. Heterogeneity in the devices could lead to slight differences in the overnight oximetric profile able to influence the diagnostic ability of the signal. Additionally, the $SpO_2$ signal contains artifacts caused by loss of contact of the pulse oximeter probe due to subject movements. Therefore, a signal pre-processing step was included to standardize the signals obtained from the different recording devices, as well as to remove motion artifacts. Signal pre-processing was different for the two main methodological approaches conducted in this research: (i) feature-engineering and (ii) deep-learning approaches.

Following a feature-engineering methodology, artifacts were first discarded from $SpO_2$ recordings by removing samples with $SpO_2$ values below 50% as well as sudden changes with a slope $\geq 4\%$ per second [15]. As the sampling frequency among the different recording devices ranged from 1 to 512 Hz, oximetry signals were re-sampled to a common sample rate of: (i) 25 Hz in Vaquerizo-Villar et al. [12, 13], as advocated by the AASM [2]; and (ii) 1 Hz for the multiscale analysis of the oximetry signal proposed in Vaquerizo-Villar et al. [14], which has been considered more appropriate for multiscale analysis approaches in previous studies [16, 17]. Finally, the resolution of the $SpO_2$ recordings was set to two decimal points (resolution of 0.01%) in order to homogenize the resolution of signals from different recording devices [18].

In the deep-learning branch, the pre-processing stage was simpler, as deep-learning approaches are able to automatically process raw data [19]. In this respect, $SpO_2$ signal pre-processing consisted on: (i) down-sampling of the $SpO_2$ recordings 1 Hz in order to homogenize the frequency [20]; and (ii) segmentation of the $SpO_2$ recordings into 20-min segments (1200 samples) prior to train the CNN-based deep-learning architecture [11].

## 3.3 Feature Engineering

Feature engineering is the conventional methodology employed to analyze biomedical signals. This methodology consists of the three following stages: (i) feature extraction; (ii) feature selection; and (iii) pattern recognition.

### *3.3.1 Feature Extraction*

As aforementioned, $SpO_2$ signals were characterized both in time and frequency domains. In this respect, different signal processing methods were applied to obtain OSA-related features from the oximetry signal [11–14].

#### 3.3.1.1 Conventional Measures in the Time Domain

In this research, conventional features were obtained from the oximetry signal using different time-domain analysis methods: oxygen desaturation index, statistical moments, and nonlinear parameters [11–14]. A description of these methods is found below.

**Oxygen desaturation index**. As mentioned in Sect. 1.4.1, oxygen desaturations are related to apneic episodes [2]. In this research, the number of oxygen desaturations of at least 3% (ODI3) from prior $SpO_2$ baseline was computed [21].

**Statistical moments**. Common first-to-fourth order statistical moments ($M1_t$–$M4_t$) were calculated to characterize the $SpO_2$ signal amplitude distribution. Accordingly, mean ($M1_t$), variance ($M2_t$), skewness ($M3_t$), and kurtosis ($M4_t$) allow to measure the central tendency, dispersion, asymmetry, and peakedness of the data, respectively.

**Nonlinear parameters**. In recent years, nonlinear methods derived from the chaos theory have shown its usefulness to extract additional information of oximetry dynamics in both adults and pediatric OSA patients [22–26]. In this respect, the following common nonlinear features have been obtained from each $SpO_2$ recording:

- Central tendency measure (*CTM*). Using second-order differences plots [27], *CTM* allows to quantify the variability of the oximetry signal associated to apneic events.

- Lempel-Ziv complexity (*LZC*). *LZC* measures changes in the complexity of the oximetry signal related to pediatric OSA severity by transforming the $SpO_2$ recordings into a two-symbol sequence and quantifying the number of different substrings in this transformed sequence [28].
- Sample entropy (*SampEn*). *SampEn* allows to quantify the irregularity of the oximetry signal by the evaluation of both prevailing and secondary patterns in the the $SpO_2$ recordings [29].

#### 3.3.1.2 Novel Oximetric Indices in the Time Domain: Detrended Fluctuation Analysis

DFA is an important tool to analyze the correlation properties of a non-stationary time series (i.e., the oximetry signal) through its multiscale analysis [30]. In this respect, DFA allows to detect changes in the correlation properties of a signal along temporal scales caused by random spikes and/or segments with a distinct local behavior [17, 31]. In Vaquerizo-Villar et al. [14], we propose DFA as a novel nonlinear analysis method to analyze the irregular fluctuations and random spikes of the $SpO_2$ signal related to apneic events. Given a time series $x(t)$, the DFA procedure comprises the following four steps [30]:

(1) Integration of the input signal. $x(t)$ is integrated using the following expression:

$$y(i) = \sum_{j=1}^{i} [x(j) - x_{mean}], \quad i = 1, \ldots, N, \tag{3.1}$$

where $x_{mean}$ and $N$ are the mean and length of the whole time series, respectively.

(2) Window-segmentation. $y(i)$ is divided into $B$ non-overlapping time windows of equal size. The length of each window (i.e., the scale) ranges from $k = 3$ to $k = 1080$, being the maximum scale (1080) one-tenth of the minimum signal length (10,800 samples = 3 h with a sampling 1 Hz), which ensures that the recording contains an adequate number of sleep cycles [32, 33].

(3) Obtaining of the local trend. A straight line least-squares fit is applied to $y(i)$ in order to obtain the local trend $y^b$ for each window ($b = 1, \ldots, B$).

(4) Obtaining of the fluctuation function. The fluctuation function, $F(k)$, is obtained for each scale using the following expression:

$$F(k) = \sqrt{\frac{1}{B} \cdot \sum_{b=1}^{B} F_b^2(k)}, \tag{3.2}$$

where $F_b^2(k)$ is the variance of the fluctuation function in each window, defined as follows:

$$F_b^2(k) = \frac{1}{k} \sum_{j=(b-1)k+1}^{bk} [(y(j) - y^b(j))^2]. \tag{3.3}$$

Steps 2–4 are iterated until $F(k)$ is obtained for each scale ($k$) of the DFA profile. A double logarithmic-scale plot was used to analyze the evolution of the DFA profile of a subject along time scales: $\log_{10} F(k)$ versus $\log_{10}(k)$. In Vaquerizo-Villar et al. [14], we obtained two different scaling regions according to the linear relationship between $\log_{10} F(k)$ and $\log_{10}(k)$: region 1 was obtained in the range of scales $0.48 \leq \log_{10}(k) \leq 1.3$ ($3 \leq k \leq 20$), whereas region 2 was obtained in the range $1.60 \leq \log_{10}(k) \leq 3.03$ ($40 \leq k \leq 1080$). In order to quantify the differences in the DFA plot associated to pediatric OSA and its severity, the following features were extracted [14, 17, 34]:

- Scaling exponents (slopes) in the straight line that fits both regions of the DFA plot ($slope_1$ and $slope_2$), which are intended to characterize the scaling behavior of the $SpO_2$ signal in each region.
- Ratio between $slope_1$ and $slope_2$ ($slope_{12}$), which characterizes the distinct scaling behavior observed in both regions.
- Coordinates of the intersection between the straight lines adjusted in regions 1 and 2 of the DFA plot ($k_{12}$ and $F(k_{12})$), which characterize the crossover point of the DFA profile [14].
- Value of the fluctuation function ($F(k_x)$) in the scale $k_x = 21$, which has the highest correlation with the AHI. $F(k_x)$ quantifies the fluctuations of the $SpO_2$ recording associated to apneic events.

#### 3.3.1.3 Conventional Frequency Domain Analysis: Power Spectral Density

Frequency domain analysis allows to measure the recurrence and duration of apneic events from OSA in children [35]. In this respect, PSD is the most used frequency domain technique to analyze the spectral components of the oximetric dynamics in the framework of pediatric OSA [23–26, 35]. PSD was computed using the Welch's approach [36] to look for differences in the oximetric recordings related to apneic events due to OSA. In order to characterize the effects of OSA in the PSDs of the $SpO_2$ recordings, Vaquerizo-Villar et al. [12] and Hornero et al. [25] determined two frequency bands of interest (BI): 0.018–0.050 Hz ($BI1$) and 0.021–0.044 Hz ($BI2$), respectively. In these bands, the highest statistically significant differences were obtained in the PSD amplitude among OSA severity groups. Once spectral bands were established, the following features were computed from the PSDs of each $SpO_2$ recording [11–14]:

- First-to-fourth order statistical moments, extracted from the full PSD ($M1_{PSD}$ –$M4_{PSD}$) and the two bands of interest: $BI1$ ($M1_{BI1}$–$M4_{BI1}$) and $BI2$ ($M1_{BI2}$ –$M4_{BI2}$). These parameters measure the central tendency, dispersion, asymmetry and peakedness of the PSD in each region, respectively.

- Relative power ($RP_{BI1}$), defined as the power ratio between $BI1$ and the whole PSD. $RP_{PSDBI1}$ is intended to reflect the effects in the band of interest caused by a higher occurrence of apneic events at these frequencies.
- Maximum amplitude ($MA_{PSD}$) and minimum amplitude ($mA_{PSD}$) of the PSD in the full spectrum ($MA_{PSD}$ and $mA_{PSD}$), $BI1$ ($MA_{BI1}$ and $MA_{BI1}$), and $BI2$ ($MA_{BI2}$ and $MA_{BI2}$). These features allow to quantify the highest and the lowest values of the PSD in these regions related to the occurrence of apneic events.
- Spectral entropy of the full PSD ($SE_{PSD}$) and $BI1$ ($SE_{BI1}$), which are irregularity parameters that measure the flatness of the PSD and its band of interest, respectively [37].
- Mobility of the PSD in the band $BI1$ ($Mb_{BI1}$), defined as the squared root of the ratio between the variance and the power of the PSD in this band. $Mb_{BI1}$ quantifies the concentration of the signal power [38].
- Median frequency of the full PSD ($MF_{PSD}$), defined as the spectral component that separates the PSD into two regions, each of them with 50% of the total power. $MF_{PSD}$ characterizes the distribution of the PSD content [37].
- Wootter's distance of the full PSD ($WD_{PSD}$), which is a disequilibrium parameter that measures the distance between the probability density function of the PSD and an uniform distribution [39].

#### 3.3.1.4 Novel Frequency Domain Features from the Oximetry Signal: Bispectrum

Bispectrum can be described as the spectral representation of the third-order cumulant (skewness) of a time series [40]. In contrast to conventional PSD, bispectrum preserves both amplitude and phase information of the spectral components of a signal. This allows bispectrum to detect phase relationships and deviations from linearity and Gaussianity in a signal [40], such as those produced in physiological signals by respiratory events [41, 42]. For these reasons, bispectral analysis is employed for the first time in Vaquerizo-Villar et al. [12] to characterize changes produced in the oximetry signal by respiratory events.

Given a deterministic and zero-mean time series $x(t)$, bispectrum can be non-parametrically estimated as follows:

$$B(f_1, f_2) = X(f_1) \cdot X(f_2) \cdot X^*(f_1 + f_2), f_1, f_2 = 0, \ldots, N, \qquad (3.4)$$

where $X(f)$ is the discrete Fourier transform of $x(t)$ and $f_1$ and $f_2$ are the frequencies of the two axes of the bispectrum. According to its symmetry conditions, bispectrum was computed in the non-redundant region ($\Omega$), which satisfies $f_1 \geq 0$, $f_2 \geq f_1$, and $f_1 + f_2 \leq f_s/2$, being $f_s$ the sample rate of $x(t)$.

Once the bispectrum was computed in this region, the following parameters were computed:

- Mean amplitude of the bispectrum ($M1_{BISP}$) [43], which allows to detect deviations of Gaussianity in a signal [44]:

$$M1_{BISP} = \frac{1}{L} \sum_{f1,f2 \in \Omega} |B(f_1, f_2)|, \tag{3.5}$$

where $L$ is the number of points of the bispectrum in $\Omega$.

- Sum of logarithmic amplitudes of the bispectrum ($H1_{BISP}$), sum of logarithmic amplitudes of diagonal elements of the bispectrum ($H2_{BISP}$), and spectral moment of first order of diagonal elements of the bispectrum ($H3_{BISP}$) [45], which are related to the moments of the bispectrum. $H1_{BISP}$, $H2_{BISP}$, and $H3_{BISP}$ quantify the non-linearity of a signal and are calculated based on the bispectral amplitudes contained in $\Omega$ and its main diagonal [40, 45]:

$$H1_{BISP} = \sum_{f_1, f_2 \in \Omega} \log(|B(f_1, f_2)|), \tag{3.6}$$

$$H2_{BISP} = \sum_{f_k \in \Omega_{diagonal}} \log(|B(f_k, f_k)|), \tag{3.7}$$

$$H3_{BISP} = \sum_{f_k \in \Omega_{diagonal}} k \cdot \log(|B(f_k, f_k)|), \tag{3.8}$$

where $\Omega_{diagonal}$ is the main diagonal of the bispectrum.

- Bispectral amplitude entropies of first ($BE1_{BISP}$) and second order ($BE2_{BISP}$), which quantify the irregularity of the bispectral amplitude of a signal [43]:

$$BE1_{BISP} = -\sum_{j \in \Omega} p_j \cdot \log(p_j), \tag{3.9}$$

where

$$p_j = \frac{|B(f_1, f_2)|}{\sum_{f1,f2 \in \Omega} |B(f_1, f_2)|}. \tag{3.10}$$

$$BE2_{BISP} = -\sum_{j \in \Omega} q_j \cdot \log(q_j), \tag{3.11}$$

where

$$q_j = \frac{|B(f_1, f_2)|^2}{\sum_{f1,f2 \in \Omega} |B(f_1, f_2)|^2}. \tag{3.12}$$

- Phase entropy ($PE_{BISP}$), which measures irregularity in the bispectral phase [43]:

$$PE_{BISP} = -\sum_{n} p(\Psi_n) \cdot \log(p(\Psi_n)), \quad n = 0, 1, \ldots, N-1, \tag{3.13}$$

where $N$ is the number of bins of the histogram and $p(\Psi_n)$ is the distribution of the phase angles [40, 46]:

$$p(\Psi_n) = \frac{1}{L} \sum_{f1, f2 \in \Omega} Ind(\phi(B(f_1, f_2)) \in \Psi_n)|, \tag{3.14}$$

where $\phi$ is the phase angle of the bispectrum and $Ind(\cdot)$ is an indicator function, whose value is 1 if $\phi$ is within the range of the bin $\Psi_n$, being $\Psi_n$ the range of histogram bins [40]:

$$\Psi_n = \left\{ \phi | -\pi + \frac{2\pi n}{N} \leq \phi < -\pi + \frac{2\pi(n+1)}{N} \right\}. \tag{3.15}$$

- Mean ($meanPa_{BISP}$) and variance ($varPa_{BISP}$) of the bispectral invariant ($P(a)$), which detect third-order time correlations or phase coupling between spectral components of a chaotic process [43]. $P(a)$ is defined as the phase of the integrated bispectrum along the straight line passing through the origin with slope $a$ [43]:

$$P(a) = \arctan\left(\frac{I_i(a)}{I_r(a)}\right), \tag{3.16}$$

where $I_r(a)$ and $I_i(a)$ are the real and imaginary components of the integrated bispectrum ($I(a)$) [43]:

$$I(a) = \int_{f_1=0^+}^{1/1+a} B(f_1, af_1)df_1 = I_r(a) + j \cdot I_i(a), \quad 0 \leq a \leq 1. \tag{3.17}$$

#### 3.3.1.5 Novel Frequency Domain Features from the Oximetry Signal: Wavelet Transform

Wavelet transform (WT) is a suitable method to analyze the spectral content of non-stationary signals [47]. In contrast to the fixed analysis window used by conventional frequency domain analysis techniques, which are based on the Short Time Fourier Transform (STFT), WT employs long windows at low frequencies and short windows at high frequencies. These multiscale analysis approach provide high frequency resolution at low frequencies and high temporal resolution at high frequencies, whereas the single scale analysis of STFT-based techniques is limited by its fixed time-frequency resolution [47]. Due to this multiresolution property, WT is applied in Vaquerizo-Villar et al. [13] as a novel technique especially designed to accurately detect $SpO_2$ desaturations elicited by apneic events, which have very low frequency components due to the slow variation nature of the oximetry signal.

WT decomposes a signal $x(t)$ onto a set of basis functions, known as wavelets [47]. Wavelets are obtained by means of the scaling and time translation of a basic wavelet prototype, the mother wavelet [47]. According to the scale and translation values, there are two main types of WT [47]: (i) continuous wavelet transform (CWT), where these time-scale parameters are continue; and (ii) discrete wavelet transform (DWT), where wavelets are computed only for power of 2 scales. In Vaquerizo-Villar et al. [13], DWT was chosen to analyze $SpO_2$ recordings, as it is less complex and computationally less expensive than CWT [47].

Figure 3.2 shows the computation process of DWT. Given a $SpO_2$ recording $x[n]$ with length $N$, DWT decomposes it using a filter-bank tree with $L = \log_2(N)$ steps [47], as shown in Fig. 3.2a. Each step consists of a half-band high-pass filter $g[n]$, the mother wavelet, and a half-band low-pass filter $h[n]$, the mirror version of $g[n]$, each of them followed by a subsampling operation with a factor two (dyadic sampling). Thereby, in the first step ($i = 1$), the $SpO_2$ signal $x[n]$ is decomposed into an approximation $A_1$ (low-pass) and a detail $D_1$ (high-pass) signal. Next, the approximation signal $A_1$ is further decomposed, thus generating another approximation signal $A_2$ and another detail $D_2$ signal. This decomposition iterates on the approximation signal until the maximum level of decomposition $i = L$. The frequency resolution of the approximation $A_i$ and detail signals $D_i$ is increased at each iteration by a factor two, while reducing their time resolution due to the dyadic sampling. At each step $i = 1, 2, \ldots, L$, $A_i$ and $D_i$ are obtained as follows:

$$A_i[k] = \sum_n A_{i-1}[n] \cdot h[2k - n], \tag{3.18}$$

$$D_i[k] = \sum_n A_{i-1}[n] \cdot g[2k - n], \tag{3.19}$$

where $A_{i-1}$ is the approximation signal in the step $i - 1$, being $A_0$ the time series $x[n]$.

In Vaquerizo-Villar et al. [13], DWT was applied to $N = 2^{13}$ (5.5 min) sample segments of the $SpO_2$ signal. Haar was chosen as the mother wavelet due to: (i) its stepped shape, which allows to detect oxygen desaturations elicited by apneic episodes; (ii) its smoothing property, which does not cause edge effects in the $SpO_2$ signal waveform.

Figure 3.2b shows an example of DWT decomposition of a $SpO_2$ segment, $x[n]$, using the Haar wavelet. It can be seen that detail coefficients from each decomposition level ($D_i$) provide information about a specific frequency range. Our DWT analysis mainly focused on $D_9$ (0.0244–0.0488 Hz), which is comprised in the frequency range of interest $BI1$ (0.018–0.050 Hz) related to the changes caused by pediatric OSA in nocturnal oximetry [13]. As it can be seen in Fig. 3.2b, $D_9$ coefficients show decreased and increased values associated to oxygen desaturations. The presence of

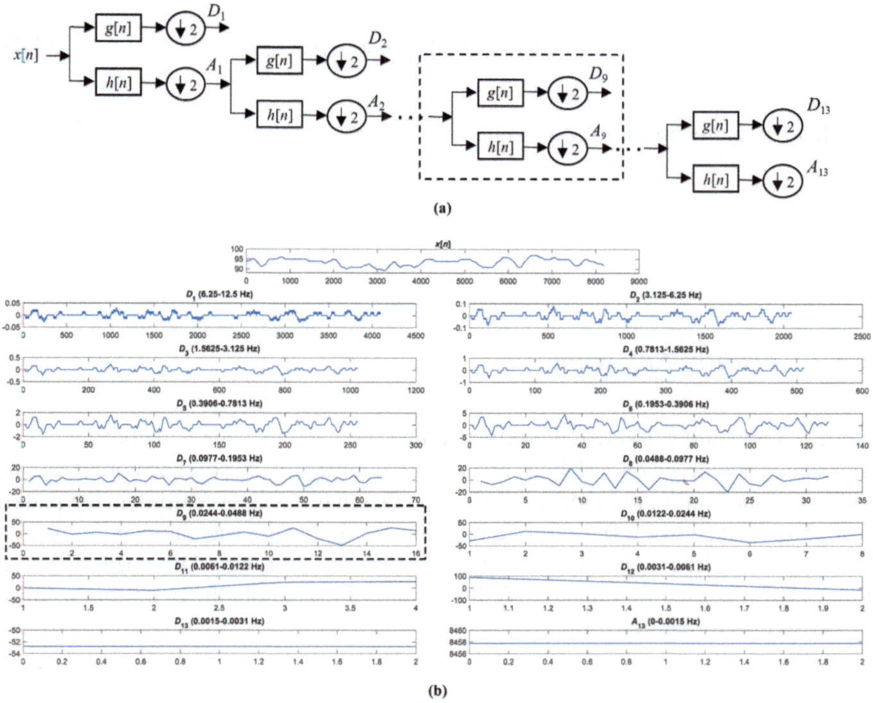


**Fig. 3.2** DWT computation. **a** Decomposition process of a signal using DWT. **b** Original $SpO_2$ signal, detail signals at each decomposition level and approximation signal at the maximum level of the decomposition. This figure has been taken from Vaquerizo-Villar et al. [13]

both negative and positive values may hinder the information contained in the DWT coefficients, such as the mean or skewness. Thereby, the following DWT-derived features were extracted from the $D_i$ coefficients in absolute value [13]:

- First-to-fourth order statistical moments of the $D_9$ coefficients ($M1_{D_9}$–$M4_{D_9}$), which are the mean ($M1_{D_9}$), variance ($M2_{D_9}$), skewness ($M3_{D_9}$), and kurtosis ($M4_{D_9}$). They measure the central tendency, dispersion, asymmetry and peakedness of the data, respectively.
- Maximum amplitude of the $D_9$ coefficients ($MA_{D_9}$), which allows to quantify the maximum amplitude reached in this frequency range.
- Energy of the $D_9$ coefficients ($En_{D_9}$), which allows to quantify the averaged quadratic amplitude of the detail signal $D_9$. It is calculated using the following expression:

$$En_{D_9} = \sum_k |D_9[k]|^2. \tag{3.20}$$

- Wavelet entropy ($WE$), which quantifies the energy distribution changes elicited in the detail levels of the oximetry signal by apneic events [48]:

$$WE = -\sum_{i=1}^{L} p_i \log(p_i), \tag{3.21}$$

where $p_i$ is the normalized wavelet energy at the detail level $i$:

$$p_i = \frac{En_{D_i}}{L} \sum_{i=1}^{L} En_{D_i}. \tag{3.22}$$

## 3.3.2 *Feature Selection*

Once features are extracted, the information contained in a $SpO_2$ recording is synthesized in a wide set or variables. Nonetheless, there may be some features that provide irrelevant and/or redundant information that may cause overfitting in the pattern recognition stage [49]. Therefore, a feature selection stage was included to prevent from this undesired effect. Specifically, the fast correlation-based filter (FCBF) algorithm [50] was used in Vaquerizo-Villar et al. [12–14] to obtain an optimum subset of OSA-related features [50].

### 3.3.2.1 Fast Correlation-Based Filter (FCBF)

FCBF is an automated feature selection algorithm that has been widely used in the pediatric OSA context [12–14, 25, 51–53]. FCBF follows a two-stage procedure to evaluate the relevance and redundancy of the variables under study. Given a feature vector $x_i$ and a dependent variable $y$, FCBF first calculates the symmetrical uncertainty ($SU$) between each single feature $x_i$ and $y$ in order to evaluate its relevancy [50]:

$$SU(x_i, y) = 2 \cdot \left( \frac{IG(x_i|y)}{H(x_i) + H(y)} \right), \quad i = 1, 2, \ldots, N, \tag{3.23}$$

where $IG(x_i|y) = H(x_i) - H(x_i|y)$, $H$ refers to the well-know Shannon's entropy, $N$ is the number of input variables [50], and $y$ is a context-dependent characteristic. Particularly, in the framework of pediatric OSA, $y$ is a vector composed of the AHI values for each pediatric subject. $SU$ can take values in the range 0–1. $SU = 0$ indicates that the two variables are totally independent, whereas $SU = 1$ indicates that it is completely possible to forecast one feature from the other. Therefore, features having a higher value of $SU$ are more relevant, as they share more information with the dependent variable [50].

Once $SU$ is obtained for all the input variables, features are sorted from the most relevant (highest $SU$) to the least relevant one (lowest $SU$). Based on this ranking, the second step consists on the computation of $SU$ between each pair of features ($SU(x_i, x_j)$), $j = 1, 2, \ldots, N$, $j \neq i$ to assess its redundancy, beginning from the most relevant ones [50]. If $SU(x_i, x_j) \geq SU(x_i, y)$, being $x_i$ more relevant than $x_j$, $x_j$ is removed from the selection process due to redundancy with $x_i$. Accordingly, an optimum subset of features is obtained with those not discarded during the redundancy analysis.

In the present doctoral thesis, the FCBF algorithm was applied in the three studies that followed a feature-engineering methodology [12–14]. In these papers, FCBF was applied to 1000 bootstrap-derived samples from the corresponding training set, as it contributes to select a more stable subset of features [54]. Then, the features selected at least 500 times (50% of the iterations) constituted the optimum subset.

### *3.3.3 Pattern Recognition*

Pattern recognition consists on the application of automated algorithms to identify the underlying behavior in the data [55]. Taking as input data the optimum subset of features from each subject obtained with FCBF, three pattern recognition algorithms have been applied in the present doctoral thesis to automatically detect pediatric OSA and its severity: (i) LR; (ii) SVM; and (iii) MLP.

#### 3.3.3.1 Logistic Regression (LR)

LR is a common method for two-class (binary) classification. It is a supervised algorithm which assigns an input feature vector $x_i$ (with $i = 1, 2, S$, being $S$ the number of instances) into one out of two mutually exclusive groups ($C_j = 1, 2$) [56]. In this thesis, LR was used to estimate the posterior probability of a given instance $x_i$ (subject) belonging to the negative OSA ($C_1$) and positive OSA ($C_2$) groups [13]. This was carried out through the logistic function [56]:

$$p(C_j|x_i) = \frac{\exp(\beta_0 + x_i\beta)}{1 + \exp(\beta_0 + x_i\beta)}, \tag{3.24}$$

where $\beta_0$ and $\beta = \beta_1, \beta_2, \ldots, \beta_N$ are the coefficients of the LR model, and $N$ is the number of input features that compose each vector $x_i$. $\beta_0$ and $\beta$ coefficients are optimized using the maximum likelihood estimator [56].

#### 3.3.3.2 Support Vector Machines (SVM)

SVM is a supervised algorithm for binary classification that finds a separating hyperplane with a decision boundary that maximizes the distance between instances belonging to different classes. The hyperplane is expressed as follows [55]:

$$y(x, w) = w^T \cdot \phi(x) + w_0, \tag{3.25}$$

where $x \in R_N$ is the input feature vector, $N$ is the number of features, $\phi(x) \in R_D$ is the feature vector in the high-dimensional transformed space ($D > N$), and $w$ denotes the weight vector, which is optimized in order to obtain a maximal margin hyperplane [55]. This optimization can be formulated using Lagrange multipliers:

$$y(x, w) = -\sum_{j \in V} \eta^j t^j K(x^j, x) + w_0, \tag{3.26}$$

where $V$ is a subset of indices $1, \ldots, S$ corresponding to the Lagrange multipliers $\eta^j$ related to the supported vectors, $S$ is the number of examples in the training group, $K(\cdot, \cdot)$ is the Kernel function, and $t^j$ are the output labels. The Lagrange multipliers $(\eta^1, \ldots, \eta^S)$ are subjected to the following constraints:

$$\sum_{j=1}^{S} \eta^j t^j = 0 \tag{3.27}$$

and

$$0 \leq \eta^j \leq C, \tag{3.28}$$

where $C$ is the regularization parameter, which controls the balance between maximizing the margin of separation between classes and minimizing the classification error.

In Vaquerizo-Villar et al. [13], SVM was applied using a linear kernel to assign every input feature vector to the groups negative OSA ($t^j = -1$) and positive OSA ($t^j = 1$), whereas the optimum value for the regularization parameter $C$ was obtained using a 10-fold cross validation procedure during the training stage.

#### 3.3.3.3 Multilayer Perceptron Neural Network (MLP)

Artificial neural networks (ANN) arised from the need to model information processing in biological systems using mathematical representations [55]. MLP is the ANN-based pattern recognition algorithm most widely used in the pediatric OSA context [57]. MLP is a feed-forward neural network with an architecture consisting on several fully-connected layers (input, hidden, and output layers) composed of basic mathematical units that imitate biological neurons, called perceptrons [55]. These units are described by a differentiable activation function $g(\cdot)$ that performs a non-linear transformation of the data, as well as by adaptive weights $w$ that connect each unit with every neuron from the subsequent layer. In this thesis, a configuration with a single hidden layer was used. Given an input feature vector $x_i$, $i = 1, \ldots, N$, being $N$ the number of features, the values of the output units $y_k$ of the MLP architecture are computed as follows:

$$y_k = g_o \left\{ \sum_{j=1}^{N_H} w_{jk} g_h \left\{ \sum_{i=1}^{N} w_{ij} x_i + b_j \right\} + b_k \right\}, \quad k = 1, \ldots, K, \tag{3.29}$$

where $g_o(\cdot)$ and $g_h(\cdot)$ are the activation functions for units in the output and hidden layer, respectively, $w_{jk}$ is the weight connecting the neuron $j$ of the hidden layer with the output unit $y_k$, $w_{ij}$ is the weight connecting the input feature $i$ with the hidden neuron $j$, $b_k$ and $b_j$ are the biases of the output and hidden layers, respectively, $K$ is the number of output units, and $N_H$ is the number of neurons in the hidden layer. In this thesis, MLP was used in Vaquerizo-Villar et al. [12] for multiclass classification (negative OSA, moderate OSA, and severe OSA), in Vaquerizo-Villar et al. [13] for binary classification (negative OSA and positive OSA), and in Vaquerizo-Villar et al. [14] for regression (AHI estimation). Accordingly, the number of output units $K$ was different in each article: $K = 3$ for three-class classification [12], $K = 2$ for binary classification [13], and $K = 1$ for AHI estimation [14].

Network weights were adjusted during training using the scale conjugate gradient algorithm with weigh-decay regularization, which allows to minimize the loss function and achieve a good generalization ability [55]. During the training process, two hyperparameters were optimized through 10-fold cross validation: the regularization parameter ($\alpha$) and $N_H$.

## 3.4 Deep Learning

Conventional feature-engineering approaches have two main limitations [58]: (i) a substantial knowledge in the specific field is required to determine, a priori, a set of relevant features that must be obtained from the input data; and (ii) these methods are limited by their low level of abstraction, which limits their capability to find relevant features in the raw data. These limitations may lead to the omission of specific features from the oximetry signal linked to the physiological perturbations of pediatric OSA. These shortcomings can be minimized by means of the application of deep-learning algorithms. As aforementioned, deep-learning methods have beaten conventional approaches in many fields [59], including the OSA context [19, 60], primarily due to their capability to automatically discover intricate patterns from the raw data using ANNs with a high number of hidden layers [60]. In this research, we have evaluated a new deep-learning model based on CNNs [11], the most widely-used deep-learning architecture in the OSA context [19, 60]. Despite the fact that CNNs were originally inspired to process image data, these architectures have proven to be the most appropriate for time series classification in many domains [61], including biomedical signal analysis [62–65]. CNNs have a multi-layer architecture whose main properties are: shared weights, sparse connections, and pooling operations [58]. The first layers of the CNN architecture identify local motifs, whereas the deeper layers detect long-term patterns occurring in different parts of the array [62]. Accordingly, a CNN architecture is applied in Vaquerizo-

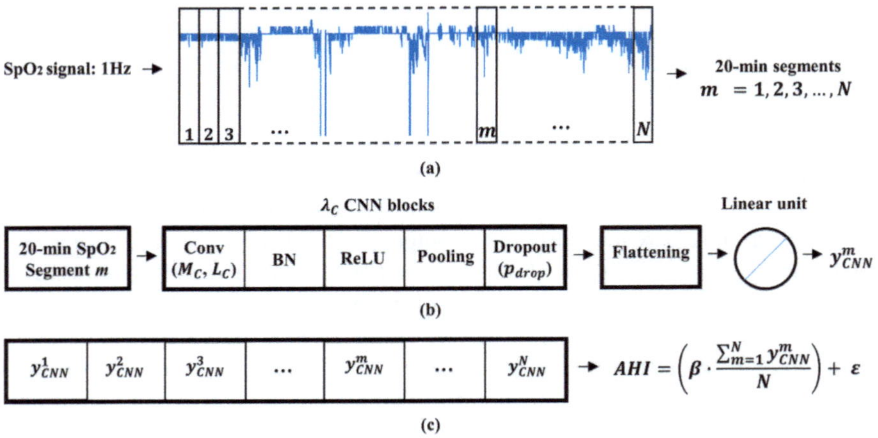


**Fig. 3.3** Overview of the proposed CNN-based deep-learning methodology. **a** Signal segmentation, **b** CNN architecture, and **c** AHI estimation. © 2021 IEEE. Reprinted, with permission, from Vaquerizo-Villar et al. [11]

Villar et al. [11] to identify oxygen desaturations (simple patterns) and clusters of desaturations (complex patterns) linked to pediatric OSA and its severity occurring in different time locations of the $SpO_2$ recording.

## *3.4.1 Proposed Convolutional Neural Network (CNN) Model*

Figure 3.3 depicts a visual overview of the CNN-based deep-learning model, which consists of three stages: (i) signal segmentation; (ii) CNN architecture; and (iii) AHI estimation.

### 3.4.1.1 Signal Segmentation

$SpO_2$ recordings were first decimated to a common sample frequency 1 Hz. As shown in Fig. 3.3a, $SpO_2$ signals were divided into 20-min non-overlapping segments (1200 samples). This segment size was chosen as it allows to detect clusters of desaturations linked to apneic events from pediatric OSA, whose duration is of at least 10-min [66]. Then, each 20-min $SpO_2$ segment used to train the CNN architecture was labelled with the respiratory event annotations scored by the technicians. In this respect, only the CHAT dataset was used for training purposes, as the UofC and BUH databases do not contain files with annotations of time location of apnea and hypopnea events. Accordingly, the output label for each 20-min segment of the CHAT dataset was obtained as the total number of apneic events (apnea plus hypopnea) associated to

3% oxygen desaturations occurring in this time window, according to the annotations provided by sleep technicians [9].

#### 3.4.1.2 CNN Architecture

Figure 3.3b shows the architecture of the proposed CNN. The input section of the CNN consists of the 20-min $SpO_2$ segments. Each segment is processed using $\lambda_C$ convolutional blocks, each one ($i = 1, \ldots, \lambda_C$) consisting of:

I. 1-D convolution. This layer extracts feature maps from the input $a_i[n]$ using $M_C$ convolutional filters, so-called kernels [58]:

$$x_i^l[n] = \sum_{k=1}^{L_C} w_k^l * a_i[n-k+1] + b_k^l, \tag{3.30}$$

where $x_i^l$ is the $l$th feature map ($l = 1, \ldots, M_C$) in the $i$th convolutional block, $L_C$ is the filter (kernel) size, $w_k^l$ and $b_k^l$ are the kernel weights and biases, respectively, and $a_0[n]$ is the input 20-min $SpO_2$ segment.

II. Batch normalization (BN). BN is used to normalize the amplitude of each feature map $x_i^l$ obtained after the 1-D convolution [58].

III. Rectified Linear Unit (ReLU). ReLU is the most common activation function for CNNs. It performs a thresholding operation in order to decide which normalized feature maps are relevant [58]:

$$f(x) = \max(0, x). \tag{3.31}$$

IV. Pooling. After the activation function, a max-pooling operation with a pool factor $K = 2$, the standard choice for CNNs, is applied to reduce dimensionality, while retaining the most significant features [58].

V. Dropout. As a final step in each convolutional block, dropout operation was included to minimize overfitting. It randomly remove connections between network elements with a probability $p_{drop}$ during the training process [58].

Following the last convolutional block ($i = \lambda_C$), a flattening layer is applied to convert the 2-D feature maps into 1-D data [58]. Finally, a linear activation unit estimates the number of apneic episodes associated to desaturations, $y_{CNN}^m$, which is the output of the network.

#### 3.4.1.3 AHI Estimation

Once the output $y_{CNN}^m$ of the CNN is obtained for each segment $m = 1, 2, \ldots, N$, the AHI of each pediatric subject can be computed. First, the mean value of all the outputs of the CNN of each patient is obtained using the following expression:

$$y_{CNN}^{avg} = \frac{\sum_{m=1}^{N} y_{CNN}^{m}}{N}, \tag{3.32}$$

where $N$ is the number of segments in the $SpO_2$ signal, which is different for each patient. Then, the final AHI of the patient is calculated using a linear regression model, as depicted in Fig. 3.3c:

$$AHI = (\beta \cdot y_{CNN}^{avg}) + \epsilon, \tag{3.33}$$

where $\epsilon$ and $\beta$ are the disturbance and interception terms of a linear regression model, which was adjusted during the optimization stage. This procedure allows to counteract the underestimation of the AHI due to [67]: (i) not all apneic events result in an oxygen desaturation, so that the CNN can not detect them; (ii) AHI estimation is performed using the total recording time, as the total sleep time is unknown using only the oximetry signal.

### 3.4.2 CNN Training and Optimization Process

The network training was performed using the following configuration: He-normal method for weights initialization [68]; adaptive moment estimation (Adam) algorithm with an initial learning rate of 0.001 for the optimization of weights and biases [69]; Huber loss with a tunable hyperparameter $\delta$ as the loss function to minimize during training [70]; batch size of 100 with a data shuffling strategy to accelerate the convergence of Adam method [58]; and 500 epochs. To speed up training and obtain a final stable set of network weights, the learning rate was decreased by a factor of 2 when the loss in the validation set did not improve for 10 consecutive epochs, and early stopping was applied after 30 epochs of non-improvement, restoring the weights that minimized the validation loss [58].

The hyperparameter optimization plays an essential role in the design of a suitable deep-learning model [58]. In this research, the following hyperparameters of the deep-learning model were optimized: the number of CNN blocks ($\lambda_C$), the number of filters ($M_C$) and the filter size ($L_C$) in each 1-D convolution, the dropout probability ($p_{drop}$), and the delta parameter of the Huber loss function ($\delta$). In order to reach an optimal solution, these hyperparameters were automatically optimized using Bayesian optimization with tree-structured Parzen estimator (BO-TPE) [71]. BO-TPE is considered a suitable strategy for optimization purposes, since it forms a probabilistic model that tries to approximate the objective function iteratively, based on past evaluation results [72].

## 3.5 Statistical Analysis

The following techniques have been employed to interpret and evaluate the results obtained with the signal processing methods developed in this doctoral thesis: (i) statistical hypothesis tests; (ii) diagnosis performance metrics; (iii) measures of agreement; and (iv) validation strategies.

### 3.5.1 *Statistical Hypothesis Tests*

Hypothesis tests are methods of statistical inference that assess whether it is possible to infer properties of a population from the results observed in a given data sample [73]. In this research, statistical hypothesis testing was first employed to evaluate the normality (Lilliefors test) and homoscedasticity (Leneve test) of the demographic variables and the features extracted from the $SpO_2$ recordings. As not all the demographic and oximetric variables passed normality and homoscedasticity tests, non-parametric tests were applied to search for statistical significant differences among the different groups under study (negative OSA/positive OSA, OSA-severity degrees, and validation groups) [73]. The Mann-Whitney *U* test was used to evaluate statistically significant differences between two groups (negative OSA/positive OSA and two validation groups), whereas the Kruskal-Wallis test was employed for comparisons in those cases with more than two groups (OSA-severity degrees and three validation groups). Depending on the number of subjects, two different *p*-values were used in this thesis to search for statistically significant differences. In Vaquerizo-Villar et al. [12], where the sample size was 298 pediatric subjects, a $p$-value $< 0.05$ was employed to evaluate statistical differences. Conversely, a $p$-value $< 0.01$ was employed in the remaining studies [11, 13, 14], where a larger sample size was used. In the case of multiple comparisons, the Bonferroni correction was used.

### 3.5.2 *Diagnostic Performance Metrics*

The usefulness of a diagnostic test can be expressed by the use of different statistical measures. The definition of these statistics is based on the number of subjects correctly and wrongly classified. In the problem of statistical classification, these measures are derived from the confusion matrix, which compares the class predicted by our test with the actual class obtained with the reference test. When the confusion matrix is intended to discern between two population groups (i.e., negative OSA and positive OSA), a binary confusion matrix is obtained, whose main elements are:

- True positives (*TP*). Number of subjects with the disease (positive subjects, according to the gold standard) that have been rightly classified by the test under study.

- False negatives (*FN*). Number of positive subjects (according to the gold standard) that have been wrongly classified as healthy or negative by the test under study.
- True negatives (*TN*). Number of negative subjects (according to the gold standard) that have been correctly classified by the test under study.
- False positives (*FP*). Number of negative subjects (according to the gold standard) that have been wrongly classified as positive by the test under study.

Based on the aforementioned elements [74], the following statistics have been calculated in this doctoral thesis [11–14]:

- Sensitivity (*Se*). Proportion of positive subjects correctly classified:

$$Se = \frac{TP}{TP + FN} \cdot 100 \tag{3.34}$$

- Specificity (*Sp*). Proportion of negative subjects correctly classified:

$$Sp = \frac{TN}{TN + FP} \cdot 100 \tag{3.35}$$

- Accuracy (*Acc*). Proportion of subjects correctly classified. This definition can also be extended for confusion matrices with more than two groups. In the case of binary classification, it is obtained as follows:

$$Acc = \frac{TP + TN}{TP + TN + FP + FN} \cdot 100 \tag{3.36}$$

- Positive predictive value (*PPV*). Proportion of subjects rightly classified among all the subjects that the test under study has assigned to the positive class:

$$PPV = \frac{TP}{TP + FP} \cdot 100 \tag{3.37}$$

- Negative predictive value (*NPV*). Proportion of subjects rightly classified among all the subjects that the test under study has assigned to the negative class:

$$NPV = \frac{TN}{TN + FN} \cdot 100 \tag{3.38}$$

- Positive likelihood ratio (*LR+*). Proportion of positive subjects rightly classified with respect to the proportion of negative subjects wrongly classified:

$$LR+ = \frac{Se}{1 - Sp} \tag{3.39}$$

$LR+$ varies between 1 and $+\infty$, being desired values close to $+\infty$.

- Negative likelihood ratio (*LR−*). Proportion of positive subjects wrongly classified with respect to the proportion of negative subjects right classified:

$$LR- = \frac{1 - Se}{Sp} \tag{3.40}$$

  $LR-$ varies between 0 and 1, being desired values close to 0.
- Area under the Receiver-Operating Characteristics (ROC) curve. The ROC curve is commonly employed to compare the yield of different diagnostic tests. A ROC plot represents the *Se* versus 1 − *Sp* curve, where *Se* and *Sp* are obtained varying the decision threshold of the test under study [75]. Once the plot has been constructed, the area under the ROC curve (AUC) can be explained as a method to assess the overall performance of a test [75]. The values of AUC are constrained to the interval between 0.5 and 1, as values below 0.5 indicate that the positiveness of the test must be changed. In this respect, a high discriminative performance is obtained when the AUC value is close to 1.

### *3.5.3 Measures of Agreement*

The agreement between the gold standard and the simplified diagnostic alternatives under study was assessed by means of the following measures:

- Cohen's kappa (kappa). kappa measures the agreement between observed and predicted classes, i.e., when considering binary or multiclass classification, without considering the agreement that may occurs by chance [76]. It is computed as:

$$kappa = \frac{p_o - p_e}{1 - p_e} \tag{3.41}$$

  where $p_o$ is the observed agreement between predicted and observed classes and $p_e$ is the probability of agreement by chance. kappa varies from −1 (total disagreement) to 1 (perfect agreement), with kappa = 0 indicating that the agreement is due completely to chance [76]. kappa was used to assess the overall agreement of our signal processing algorithms to establish pediatric OSA and its severity in Vaquerizo-Villar et al. [11, 12, 14].
- Root mean square error (RMSE). RMSE is an estimate of concordance between predicted and observed continuous variables, thus being useful to assess the performance of regression methods. In this thesis, RMSE was used to measure the agreement between the AHI predicted by our signal processing algorithms ($AHI_{pred}$) and the actual AHI from PSG ($AHI_{PSG}$) [11]. Given $AHI_{pred}$ and $AHI_{PSG}$, RMSE is calculated as:

$$RMSE = \sqrt{\frac{\sum_{n=1}^{N}(AHI_{pred}(n) - AHI_{PSG}(n))^2}{N}} \tag{3.42}$$

where $N$ is the number of instances (subjects).

- Intra-class correlation coefficient (ICC). ICC is another index commonly used to evaluate the performance of a regression algorithm. In contrast to conventional correlation coefficients, ICC considers systematic error when assessing agreement. Depending on its specific purpose and the statistical model assumed, there exist several definitions of ICC [77]. In this research [11, 14], the purpose is to measure agreement between $AHI_{pred}$ and $AHI_{PSG}$, without making any ANOVA assumptions or using replicated measurements. For this case, the following definition is recommended [77]:

$$ICC = \frac{MS_S - MS_E}{MS_S - (N_O - 1) \cdot MS_E + N_O \cdot (MS_T - MS_E)/N} \tag{3.43}$$

  where $N_O$ is the number of observers, $N$ is the number of subjects considered, $MS_S$ is the subjects mean square, $MS_E$ is the error mean square, and $MS_T$ is the observers mean square. The values of ICC ranges from 0 (no agreement) to 1 (total agreement) [78].

### 3.5.4 Validation Strategies

Several validation methodologies have been applied with the aim to maximize the generalization of the results obtained during the compendium of publications. For the purpose of minimizing potential overfitting, a dataset must be divided in as many subsets as the number of degrees of freedom (number of optimization steps) needed to be fitted for a specific problem [54]. In this respect, a hold-out strategy (training-test, training-validation-test, etc.) was employed in all the articles of this compendium to properly validate their results [11–14]. When the sample size was not large enough to handle all the optimizations required, two additional validation techniques were used: stratified K-fold cross-validation and bootstrapping.

**Hold-out validation** The natural way to correctly validate the results of a given model is to use a different set to optimize each stage of the proposed methodology [54]. In the most simple case, a training group is employed to fit the model parameters and an independent test group, also called hold-out set, is used to estimate its performance [54, 55]. As previously mentioned, the dataset should be divided into more than two sets when model fitting requires more than one optimization step [54]. In this respect, a common strategy consists of using three subsets: training, used to adjust model parameters; validation, used to adjust model hyperparameters; and test, used for independent diagnostic performance assessment. When it is not possible to use a different set to optimize each stage because of the available amount of data, hold-out must be combined with additional validation strategies, such as K-fold cross-validation and bootstrapping. In this research, hold-out was used with two subsets in Vaquerizo-Villar et al. [13, 14], and with three subsets in Vaquerizo-Villar et al. [11, 12].

**Stratified K-fold cross-validation** Stratified K-fold cross-validation is another common validation approach. This technique randomly divides the data into K folds, keeping the proportion of instances pertaining to the groups under study. Then, K-1 folds are employed for model fitting (training), whereas the remaining fold is used for model evaluation (test). This procedure is repeated K times, so that each subset was considered once as the test group [79]. Once the model has been tested using all the instances, diagnostic ability statistics can be averaged across the different folds. This may lead to more generalizable models, at the cost of increasing the computational cost [54]. This validation approach was used for hyperparameters optimization in Vaquerizo-Villar et al. [12–14], and for diagnostic performance assessment in Vaquerizo-Villar et al. [13].

**Bootstrapping** A bootstrap procedure can also be used for validation purposes when the data set is small [54]. As stated in Sect. 3.3.2, FCBF was applied along with a bootstrap procedure to find a more generalizable optimum set of features [49]. Given a feature set $x$ of $S$ instances (subjects), $x = x_1, x_2, \ldots, x_S, x_b (b = 1, 2, \ldots, B)$ new sets (bootstrap replicates) of size $S$ are built by resampling with replacement from the initial set, following a uniform distribution. In this doctoral thesis [12–14], the FCBF algorithm was applied to each of these $B = 1000$ bootstrap samples, obtaining optimum subsets composed of those features selected in more than 500 replicates (see Sect. 3.3.2).

## References


1. Iber C, Ancoli-Israel S, Chesson A, Quan SF (2007) The AASM manual for the scoring of sleep and associated events: rules, terminology and technical specification. J Clin Sleep Med 3(7):752
2. Berry RB, Budhiraja R, Gottlieb DJ, Gozal D, Iber C, Kapur VK, Marcus CL, Mehra R, Parthasarathy S, Quan SF, Redline S, Strohl KP, Davidson Ward SL, Tangredi MM (2012) Rules for scoring respiratory events in sleep: update of the 2007 AASM manual for the scoring of sleep and associated events: deliberations of the sleep apnea definitions task force of the American Academy of Sleep Medicine. J Clin Sleep Med 8(5):597
3. Marcus CL, Brooks LJ, Ward SD, Draper KA, Gozal D, Halbower AC, Jones J, Lehmann C, Schechter MS, Sheldon S, Shiffman RN, Spruyt K (2012) Diagnosis and management of childhood obstructive sleep apnea syndrome. Pediatrics 130(3):e714–e755
4. Alonso-Álvarez ML, Canet T, Cubell-Alarco M, Estivill E, Fernández E, Gozal D, Jurado-Luque MJ, Lluch-Roselló MA, Martínez-Pérez F, Merino-Andren M, Pin-Arboledas G, Roure N, Sanmartí FX, Sans-Capdevila O, Segarra-Isern J (2011) Documento de consenso del síndrome de apneas-hipopneas durante el sueño en niños. Arch Bronconeumol 47(Supl 5):2–18
5. Church GD (2012) The role of polysomnography in diagnosing and treating obstructive sleep apnea in pediatric patients. Curr Prob Pediatr Adolesc Health Care 42(1):22–25

6. Tan H-L, Gozal D, Ramirez HM, Bandla HPR, Kheirandish-Gozal L (2014) Overnight polysomnography versus respiratory polygraphy in the diagnosis of pediatric obstructive sleep apnea. Sleep 37(2):255–260
7. Hunter SJ, Gozal D, Smith DL, Philby MF, Kaylegian J, Kheirandish-Gozal L (2016) Effect of sleep-disordered breathing severity on cognitive performance measures in a large community cohort of young school-aged children. Am J Respir Crit Care Med 194(6):739–747
8. Kaditis A, Kheirandish-Gozal L, Gozal D (2016) Pediatric OSAS: oximetry can provide answers when polysomnography is not available. Sleep Med Rev 27:96–105
9. Marcus CL, Moore RH, Rosen CL, Giordani B, Garetz SL, Taylor HG, Mitchell RB, Amin R, Katz ES, Arens R, Paruthi S, Muzumdar H, Gozal D, Thomas NH, Ware J, Beebe D, Snyder K, Elden L, Sprecher RC, Willging P, Jones D, Bent JP, Hoban T, Chervin RD, Ellenberg SS, Redline S (2013) A randomized trial of adenotonsillectomy for childhood sleep apnea. New Engl J Med 368(25):2366–2376
10. Redline S, Amin R, Beebe D, Chervin RD, Garetz SL, Giordani B, Marcus CL, Moore RH, Rosen CL, Arens R, Gozal D, Katz ES, Mitchell RB, Muzumdar H, Taylor H, Thomas N, Ellenberg S (2011) The childhood adenotonsillectomy trial (CHAT): rationale, design, and challenges of a randomized controlled trial evaluating a standard surgical procedure in a pediatric population. Sleep 34(11):1509–1517
11. Vaquerizo-Villar F, Alvarez D, Kheirandish-Gozal L, Gutierrez-Tobal GC, Barroso-Garcia V, Santamaria-Vazquez E, Del Campo F, Gozal D, Hornero R (2021) A convolutional neural network architecture to enhance oximetry ability to diagnose pediatric obstructive sleep apnea. IEEE J Biomed Health Inform 25(8), 2906–2916
12. Vaquerizo-Villar F, Álvarez D, Kheirandish-Gozal L, Gutiérrez-Tobal GC, Barroso-García V, Crespo A, del Campo F, Gozal D, Hornero R (2018) Utility of bispectrum in the screening of pediatric sleep apnea-hypopnea syndrome using oximetry recordings. Comput Methods Programs Biomed 156:141–149
13. Vaquerizo-Villar F, Álvarez D, Kheirandish-Gozal L, Gutiérrez-Tobal GC, Barroso-García V, Crespo A, del Campo F, Gozal D, Hornero R (2018) Wavelet analysis of oximetry recordings to assist in the automated detection of moderate-to-severe pediatric sleep apnea-hypopnea syndrome. PloS One 13(12):e0208502
14. Vaquerizo-Villar F, Álvarez D, Kheirandish-Gozal L, Gutiérrez-Tobal GC, Barroso-García V, Crespo A, Del Campo F, Gozal D, Hornero R (2018) Detrended fluctuation analysis of the oximetry signal to assist in paediatr sleep apnoea-hypopnoea syndrome diagnosis. Physiol Measur 39(11):114006
15. Magalang UJ, Dmochowski J, Veeramachaneni S, Draw A, Mador J, El-Solh A, Grant BJB (2003) Prediction of the apnea-hypopnea index from overnight pulse oximetry. CHEST J 124(5):1694–1701
16. Crespo A, Álvarez D, Gutiérrez-Tobal GC, Vaquerizo-Villar F, Barroso-García V, Alonso-Álvarez ML, Terán-Santos J, Hornero R, del Campo F (2017) Multiscale entropy analysis of unattended oximetric recordings to assist in the screening of paediatric sleep apnoea at home. Entropy 19(6):284
17. Hua C-C, Yu C-C (2017) Detrended fluctuation analysis of oxyhemoglobin saturation by pulse oximetry in sleep apnea syndrome. J Med Biol Eng 37(6):791–799
18. Garde A, Karlen W, Dehkordi P, Ansermino JM, Dumont GA (2014b) Oxygen saturation resolution influences regularity measurements. In: 36th Annual international conference of the IEEE in engineering in medicine and biology society (EMBC), pp 2257–2260
19. Mostafa SS, Mendonça F, Ravelo-García G (2019) A systematic review of detecting sleep apnea using deep learning. Sensors 19(22):4934
20. Mostafa SS., Mendonça F, Morgado-Dias F, Ravelo-García A (2017) $SpO_2$ based sleep apnea detection using deep learning. In: 2017 IEEE 21st International conference on intelligent engineering systems (INES). IEEE, pp 91–96
21. Taha BH, Dempsey JA, Weber SM, Badr MS, Skatrud JB, Young TB, Jacques AJ, Seow K (1997) Automated detection and classification of sleep-disordered breathing from conventional polysomnography data. Sleep 20(11):991–1001

22. Alvarez D, Hornero R, Marcos JV, Wessel N, Penzel T, Glos M, Del Campo F (2013) Assessment of feature selection and classification approaches to enhance information from overnight oximetry in the context of apnea diagnosis. Int J Neural Syst 23(05):1350020
23. Garde A, Dehkordi P, Karlen W, Wensley D, Ansermino JM, Dumont GA (2014) Development of a screening tool for sleep disordered breathing in children using the phone oximeter. PloS One 9(11):e112959
24. Álvarez D, Alonso-Álvarez ML, Gutiérrez-Tobal GC, Crespo A, Kheirandish-Gozal L, Gozal D, Terán-Santos J, Campo FD (2017) Automated screening of children with obstructive sleep apnea using nocturnal oximetry: an alternative to respiratory polygraphy in unattended settings. J Clin Sleep Med 13(5):7–11
25. Hornero R, Kheirandish-Gozal L, Gutiérrez-Tobal GC, Philby MF, Alonso-Álvarez ML, Álvarez D, Dayyat EA, Xu Z, Huang Y-S, Tamae Kakazu M, Li AM, Van Eyck A, Brockmann PE, Ehsan Z, Simakajornboon N, Kaditis AG, Vaquerizo-Villar F, Crespo Sedano A, Sans Capdevila O, von Lukowicz M, Terán-Santos J, Del Campo F, Poets CF, Ferreira R, Bertran K, Zhang Y, Schuen J, Verhulst S, Gozal D (2017) Nocturnal oximetry-based evaluation of habitually snoring children. Am J Respir Crit Care Med 196(12):1591–1598
26. Crespo A, Álvarez D, Kheirandish-Gozal L, Gutiérrez-Tobal GC, Cerezo-Hernández A, Gozal D, Hornero R, Del Campo F (2018) Assessment of oximetry-based statistical classifiers as simplified screening tools in the management of childhood obstructive sleep apnea. Sleep Breathing 22(4):1063–1073
27. Cohen ME, Hudson DL, Deedwania PC (1996) Applying continuous chaotic modeling to cardiac signal analysis. IEEE Eng Med Biol Mag 15(5):97–102
28. Lempel A, Ziv J (1976) On the complexity of finite sequences. IEEE Trans Inf Theor 22(1):75–81
29. Richman JS, Moorman JR (2000) Physiological time-series analysis using approximate entropy and sample entropy. Am J Physiol-Heart Circulatory Physiol 278(6):H2039–H2049
30. Peng C-K, Buldyrev SV, Havlin S, Simons M, Stanley HE, Goldberger AL (1994) Mosaic organization of DNA nucleotides. Phys Rev E 49(2):1685
31. Chen Z, Ivanov PC, Hu K, Stanley HE (2002) Effect of nonstationarities on detrended fluctuation analysis. Phys Rev E Stat Phys Plasmas Fluids Relat Interdiscip Top 65(4):15
32. Davis KF, Parker KP, Montgomery GL (2004) Sleep in infants and young children: part one: normal sleep. J Pediatr Health Care 18(2):65–71
33. Kapur VK, Auckley DH, Chowdhuri S, Kuhlmann DC, Mehra R, Ramar K, Harrod CG (2017) Clinical practice guideline for diagnostic testing for adult obstructive sleep apnea: an American academy of sleep medicine clinical practice guideline. J Clin Sleep Med 13(3):479–504
34. Penzel T, Kantelhardt J, Grote L, Peter J, Bunde A (2003) Comparison of detrended fluctuation analysis and spectral analysis for heart rate variability in sleep and sleep apnea. IEEE Trans Biomed Eng 50(10):1143–1151
35. del Campo F, Crespo A, Cerezo-Hernández A, Gutiérrez-Tobal GC, Hornero R, Álvarez D (2018) Oximetry use in obstructive sleep apnea. Expert Rev Respir Med 12(8):665–681
36. Welch PD (1967) The use of fast Fourier transform for the estimation of power spectra: a method based on time averaging over short, modified periodograms. IEEE Trans Audio Electroacoust 15(2):70–73
37. Poza J, Hornero R, Abásolo D, Fernández A (2007) Extraction of spectral based measures from MEG background oscillations in Alzheimer's disease. Med Eng Phys 29(10):1073–1083
38. Blanco-Velasco M, Cruz-Roldan F, Godino-Llorente JI, Barner KE (2010) Nonlinear trend estimation of the ventricular repolarization segment for T-wave alternans detection. IEEE Trans Bio-med Eng 57(10):2402–2412
39. Martin M, Plastino A, Rosso O (2003) Statistical complexity and disequilibrium. Phys Lett A 311(2–3):126–132
40. Chua K, Chandran V, Acharya UR, Min C (2010) Application of higher order statistics/spectra in biomedical signals: a review. Med Eng Phys 32(7):679–689
41. Atri R, Mohebbi M (2015) Obstructive sleep apnea detection using spectrum and bispectrum analysis of single-lead ECG signal. Physiol Measur 36(9):1963

42. Tagluk ME, Sezgin N (2011) A new approach for estimation of obstructive sleep apnea syndrome. Expert Syst Appl 38(5):5346–5351
43. Chua KC, Chandran V, Acharya U, Lim C (2008) Cardiac state diagnosis using higher order spectra of heart rate variability. J Med Eng Technol 32(2):145–155
44. Ning T, Bronzino J (1990) Autoregressive and bispectral analysis techniques: EEG applications. IEEE Eng Med Biol Mag 9(1):47–50
45. Zhou S-M, Gan JQ, Sepulveda F (2008) Classifying mental tasks based on features of higher-order statistics from EEG signals in brain-computer interface. Inf Sci 178(6):1629–1640
46. Doane DP (1976) Aesthetic frequency classifications. Am Stat 30(4):181–183
47. Rioul O, Vetterli M (1991) Wavelets and signal processing. IEEE Sig Process Mag 8(4):14–38
48. Rosso OA, Blanco S, Yordanova J, Kolev V, Figliola A, Schürmann M, Başar E (2001) Wavelet entropy: a new tool for analysis of short duration brain electrical signals. J Neurosci Methods 105(1):65–75
49. Guyon I (2003) An introduction to variable and feature selection. J Mach Learn Res 3:1157–1182
50. Yu L, Liu H (2004) Efficient feature selection via analysis of relevance and redundancy. J Mach Learn Res 5:1205–1224
51. Barroso-García V, Gutiérrez-Tobal GC, Kheirandish-Gozal L, Álvarez D, Vaquerizo-Villar F, del Campo F, Gozal D, Hornero R (2020) Usefulness of recurrence plots from airflow recordings to aid in paediatric sleep apnoea diagnosis. Comput Methods Programs Biomed 183:105083
52. Barroso-García V, Gutiérrez-Tobal GC, Kheirandish-Gozal L, Vaquerizo-Villar F, Álvarez D, del Campo F, Gozal D, Hornero R (2021) Bispectral analysis of overnight airflow to improve the pediatric sleep apnea diagnosis. Comput Biol Med 129:104167
53. Jiménez-García J, Gutiérrez-Tobal GC, García M, Kheirandish-Gozal L, Martín-Montero A, Álvarez D, del Campo F, Gozal D, Hornero R (2020) Assessment of airflow and oximetry signals to detect pediatric sleep apnea-hypopnea syndrome using AdaBoost. Entropy 22(6):670
54. Witten I, Frank E, Hall M (2011) Data mining. Morgan Kaufmann
55. Bishop CM (2006) Pattern recognition and machine learning. Springer, Berlin
56. Hosmer D, Lemeshow S (2004) Applied logistic regression. Wiley
57. Gutiérrez-Tobal GC, Álvarez D, Kheirandish-Gozal L, Del Campo F, Gozal D, Hornero R (2022) Reliability of machine learning to diagnose pediatric obstructive sleep apnea: systematic review and meta-analysis. Pediatr Pulmonol 57(8):1931–1943
58. Goodfellow I, Bengio Y, Courville A, Bengio Y (2016) Deep learning, vol 1. MIT Press, Cambridge
59. LeCun Y, Bengio Y, Hinton G (2015) Deep learning. Nature 521(7553):436–444
60. Faust O, Razaghi H, Barika R, Ciaccio EJ, Acharya UR (2019) A review of automated sleep stage scoring based on physiological signals for the new millennia. Comput Methods Programs Biomed 176:81–91
61. Ismail Fawaz H, Forestier G, Weber J, Idoumghar L, Muller PA (2019) Deep learning for time series classification: a review. Data Min Knowl Disc 33(4):917–963
62. Ebrahimi Z, Loni M, Daneshtalab M, Gharehbaghi A (2020) A review on deep learning methods for ECG arrhythmia classification. Expert Syst Appl X 7:100033
63. Faust O, Hagiwara Y, Hong TJ, Lih OS, Acharya UR (2018) Deep learning for healthcare applications based on physiological signals: a review. Comput Methods Programs Biomed 161:1–13
64. Murat F, Yildirim O, Talo M, Baloglu UB, Demir Y, Acharya UR (2020) Application of deep learning techniques for heartbeats detection using ECG signals-analysis and review. Comput Biol Med 103726
65. Roy Y, Banville H, Albuquerque I, Gramfort A, Falk TH, Faubert J (2019) Deep learning-based electroencephalography analysis: a systematic review. J Neural Eng 16(5):051001
66. Brouillette RT, Morielli A, Leimanis A, Waters KA, Luciano R, Ducharme FM (2000) Nocturnal pulse oximetry as an abbreviated testing modality for pediatric obstructive sleep apnea. Pediatrics 105(2):405–412

67. Deviaene M, Testelmans D, Buyse B, Borzée P, Van Huffel S, Varon C (2018) Automatic screening of sleep apnea patients based on the $SpO_2$ signal. IEEE J Biomed Health Inf 23(2):607–617
68. He K, Zhang X, Ren S, Sun J (2015) Delving deep into rectifiers: surpassing human-level performance on ImageNet classification. In: Proceedings of the IEEE international conference on computer vision, pp 1026–1034
69. Kingma DP, Ba J (2014) Adam: a method for stochastic optimization. arXiv preprint arXiv:1412.6980
70. Huber PJ (1964) Robust estimation of a location parameter. Annals Math Stat 35(1):73–101
71. Bergstra J, Bardenet R, Bengio Y, Kégl B (2011) Algorithms for hyper-parameter optimization. In: 25th Annual conference on neural information processing systems (NIPS 2011), pp 1–9
72. Snoek J, Larochelle H, Adams RP (2012) Practical Bayesian optimization of machine learning algorithms. arXiv preprint arXiv:1206.2944
73. Jobson JD (2012) Applied multivariate data analysis: regression and experimental design. Springer Science & Business Media
74. Flemons WW, Littner MR (2003) Measuring agreement between diagnostic devices. Chest 124(4):1535–1542
75. Zweig MH, Campbell G (1993) Receiver-operating characteristic (ROC) plots: a fundamental evaluation tool in clinical medicine. Clin Chem 39(4):561–577
76. Cohen J (1960) A coefficient of agreement for nominal scales. Educ Psychol Measur 20(1):37–46
77. Chen C-C, Barnhart HX (2008) Comparison of ICC and CCC for assessing agreement for data without and with replications. Comput Stat Data Anal 53(2):554–564
78. Weir JP (2005) Quantifying test-retest reliability using the intraclass correlation coefficient and the SEM. J Strength Conditioning Res 19(1):231–240
79. Steyerberg EW, Vergouwe Y (2014) Towards better clinical prediction models: seven steps for development and an ABCD for validation. Eur Heart J 35(29):1925–1931

# Chapter 4
# Results

This chapter presents the main outcomes obtained in this doctoral thesis. These results are organized according to the statements that compose the global hypothesis of Sect. 2.1, thus having almost a directly connection with the papers included in the doctoral thesis.

## 4.1 Application of Novel Feature-Extraction Algorithms

As explained in Sect. 2.3, three novel feature-extraction algorithms were applied in this research to obtain features able to provide additional information from the oximetry signal linked to apneic events due to pediatric OSA: bispectrum, wavelet analysis, and DFA.

### *4.1.1 Bispectral Analysis*

The complementarity of bispectrum to conventional approaches in the screening of childhood OSA using $SpO_2$ recordings was evaluated in Vaquerizo-Villar et al. [1]. For this purpose, up to 22 features were obtained from a database of 298 pediatric patients (i.e., the initial version of the UofC database): ODI3, 3 anthropometric variables, 9 features from the PSD, and 9 bispectral parameters. Then, the FCBF algorithm was applied to select an optimum subset of OSA-related features. Finally, a MLP neural network was trained with the selected variables to detect pediatric OSA severity degrees. According to this feature-engineering methodology, the database was split into three sets: (*i*) feature optimization set (74 subjects, 25%), employed to optimize the feature extraction and selection stages; (*ii*) training set (149 subjects, 50%), employed to optimize the hyperparameters of the MLP classifier as well as to



F. Vaquerizo Villar, *Automated Analysis of the Oximetry Signal to Simplify the Diagnosis of Pediatric Sleep Apnea*, Springer Theses,
https://doi.org/10.1007/978-3-031-32832-9_4

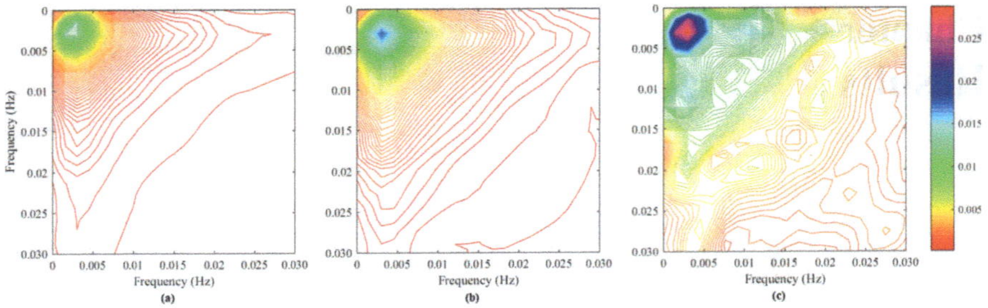


**Fig. 4.1** Averaged amplitude in absolute value of the bispectrum for the three OSA severity groups. **a** negative OSA, **b** moderate OSA, and **c** severe OSA. Reprinted from Vaquerizo-Villar et al. [1] ©(2018) with permission from Elsevier

train the MLP model; (*iii*) test set (75 subjects, 25%), used to examine the diagnostic ability of the proposed methodology in an independent set.

Figure 4.1 shows the averaged amplitude in absolute value of the bispectrum in the feature optimization set for the following OSA severity groups: negative OSA (AHI <5 e/h), moderate OSA ($5 \leq$ AHI $<$ 10 e/h), and severe OSA (AHI $\geq$ 10 e/h). As shown, a higher bispectral amplitude is appreciated in the frequency range 0–0.03 Hz as the OSA severity increases. Table 4.1 shows the median and interquartile range of the extracted features in the feature optimization set for negative OSA, moderate OSA, and severe OSA groups, along with their corresponding *p*-values, obtained by means of the Kruskal-Wallis test. ODI3, 6 out of 9 PSD-derived parameters ($M1_{BI1}$, $M2_{BI1}$, $MA_{BI1}$, $mA_{BI1}$, $SE_{BI1}$, and $Mb_{BI1}$), and 4 out of 9 bispectrum-derived features ($M1_{BISP}$, $H1_{BISP}$, $H2_{BISP}$, and $H3_{BISP}$) showed statistically significant differences ($p$-value $< 0.05$ after Bonferroni correction). Specifically, higher values were obtained in these variables as pediatric OSA severity increased.

As the goal is to assess the complementarity of bispectrum with respect to conventional approaches, two different feature sets were composed:

- $\text{set}_{\text{nobis}}$: composed of all but bispectrum features (ODI3, anthropometric variables, and PSD-derived features).
- $\text{set}_{\text{bis}}$: consisting of all the extracted features.

FCBF was applied to these two feature sets ($\text{set}_{\text{nobis}}$ and $\text{set}_{\text{bis}}$) in the feature optimization set, obtaining two optimum feature subsets ($\text{subset}_{\text{nobis}}$ and $\text{subset}_{\text{bis}}$), each one composed of those features selected in at least 50% of the 1000 bootstrap replicates (500 times) :

- $\text{subset}_{\text{nobis}}$ was composed of ODI3, the three anthropometric features (Age, Sex, and BMI), and five features from the PSD ($M1_{BI1}$, $M2_{BI1}$, $MA_{BI1}$, $RP_{BI1}$, and $Mb_{BI1}$).
- $\text{subset}_{\text{bis}}$ was composed of ODI3, two anthropometric features (Age, and Sex), five PSD-derived features ($M1_{BI1}$, $M2_{BI1}$, $MA_{BI1}$, $RP_{BI1}$, and $Mb_{BI1}$), and two bispectral features ($M1_{BISP}$, and $meanPa_{BISP}$).

**Table 4.1** Feature values for the OSA severity groups (median [interquartile range]) in the feature optimization set

| Features | Negative OSA | Moderate OSA | Severe OSA | $p$-value* |
|---|---|---|---|---|
| Age | 7.0 [4.8,11.0] | 4.0 [2.0,8.0] | 6.0 [3.0,9.0] | 0.24 |
| Sex | – | – | – | 0.52 |
| BMI($10^1$) | 1.82 [1.68,2.26] | 1.9 [1.62,2.30] | 1.77 [1.54,2.23] | 0.76 |
| ODI3 | 2.08 [0.77,3.93] | 5.82 [3.79,9.28] | 8.72 [7.23,19.65] | <0.05 |
| $M1_{BI1}$ | 2.95 [1.93,4.23] | 5.67 [5.10,8.58] | 14.73 [7.51,28.45] | <0.05 |
| $M2_{BI1}$ ($10^1$) | 0.18 [0.05,0.32] | 1.11 [0.56,1.54] | 4.07 [1.34,14.39] | <0.05 |
| $M3_{BI1}$ | 0.48 [0.29,0.75] | 0.68 [0.33,1.04] | 0.54 [0.32,0.83] | 0.54 |
| $M4_{BI1}$ | 2.10 [1.79,2.73] | 2.33 [1.87,3.06] | 2.20 [1.60,2.70] | 0.65 |
| $RP_{BI1}$ | 0.31 [0.25,0.35] | 0.24 [0.21,0.34] | 0.34 [0.30,0.39] | 0.05 |
| $MA_{BI1}$ | 0.60 [0.33,0.87] | 1.27 [1.00,1.54] | 2.92 [1.68,4.59] | <0.05 |
| $mA_{BI1}$ | 1.21 [0.84,2.12] | 2.30 [1.91,2.88] | 3.46 [2.56,8.59] | <0.05 |
| $SE_{BI1}$ | 4.33 [4.26,4.37] | 4.30 [4.24,4.32] | 4.20 [4.12,4.30] | <0.05 |
| $Mb_{BI1}$ | 0.17 [0.11,0.20] | 0.28 [0.23,0.32] | 0.42 [0.25,0.56] | <0.05 |
| $M1_{BISP}$ ($10^{-1}$) | 0.05 [0.02,0.21] | 0.23 [0.09,0.63] | 0.49 [0.17,1.22] | <0.05 |
| $H1_{BISP}$ ($10^8$) | −5.78 [−5.90,−5.57] | −5.59 [−5.73,−5.45] | −5.40 [−5.58,−5.28] | <0.05 |
| $H2_{BISP}$ ($10^5$) | −1.38 [−1.41,−1.33] | −1.32 [−1.36,−1.29] | −1.28 [−1.32, −1.25] | <0.05 |
| $H3_{BISP}$ ($10^8$) | −2.96 [−3.02,−2.86] | −2.87 [−2.93,−2.79] | −2.77 [−2.86,−2.71] | <0.05 |
| $BE1_{BISP}$ | 8.51 [7.88,9.43] | 8.17 [7.71,8.71] | 8.60 [8.11,8.79] | 0.34 |
| $BE2_{BISP}$ | 6.08 [4.88,6.62] | 5.26 [4.43,5.58] | 6.70 [5.68,7.07] | 0.13 |
| $PE_{BISP}$ | 2.14 [2.08,2.15] | 2.12 [2.05,2.14] | 2.11 [2.08,2.13] | 0.64 |
| $meanPa_{BISP}$ ($10^{-2}$) | −1.60 [−3.19,0.81] | 0.89 [−0.47,2.21] | 0.92 [−1.96,3.20] | 0.29 |
| $varPa_{BISP}$ | 0.38 [0.26,0.43] | 0.34 [0.22,0.37] | 0.33 [0.22,0.44] | 0.87 |


*$p$-values obtained from the Kruskal-Wallis test after Bonferroni correction, OSA = obstructive sleep apnea

Taking as input these optimum subsets of features ($subset_{nobis}$ and $subset_{bis}$), two MLP networks were trained and optimized using the training set ($MLP_{nobis}$ and $MLP_{bis}$). Network hyperparameters, the regularization parameter ($\alpha$) and the number of hidden units ($N_H$), were varied from $\alpha = 0$ up to $\alpha = 5$ (step of 1) and from $N_H = 2$ up to $N_H = 50$ (step of 1), respectively. The maximum value of kappa, obtained using ten-fold cross-validation, determined the optimum values of $\alpha$ and $N_H$ in each case. Thereby, $\alpha = 1$ and $N_H = 3$ were chosen for $MLP_{nobis}$ and $\alpha = 2$ and $N_H = 4$ were chosen for $MLP_{bis}$, as the pairs that reached the highest kappa.

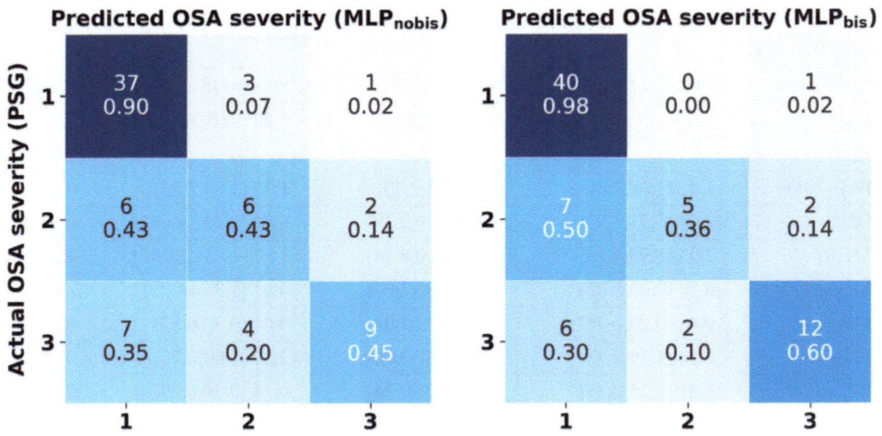


**Fig. 4.2** Confusion matrices of $MLP_{nobis}$ and $MLP_{bis}$ in the test set. 1: negative OSA, 2: moderate OSA; 3: severe OSA

**Table 4.2** Diagnostic ability of the $MLP_{nobis}$ and $MLP_{bis}$ models in the test set for the AHI cutoffs=5 e/h and 10 e/h

| | AHI cutoff | Se | Sp | PPV | NPV | LR+ | LR− | Acc | kappa |
|---|---|---|---|---|---|---|---|---|---|
| $MLP_{nobis}$ | **5 e/h** | 61.8 | 90.2 | 84.0 | 74.0 | 6.33 | 0.42 | 77.3 | 0.45 |
| | **10 e/h** | 45.0 | 94.5 | 75.0 | 82.5 | 8.25 | 0.58 | 81.3 | |
| $MLP_{bis}$ | **5 e/h** | 61.8 | 97.6 | 95.5 | 75.5 | 25.32 | 0.39 | 81.3 | 0.56 |
| | **10 e/h** | 60.0 | 94.5 | 80.0 | 86.7 | 11.00 | 0.42 | 85.3 | |


AHI = apnea-hypopnea index, Se = sensitivity (%), Sp = specificity (%), PPV = positive predictive value (%), NPV = negative predictive value (%), LR+ = positive likelihood ratio, LR− = negative likelihood ratio, Acc = accuracy (%), kappa=Cohen's kappa index

Then, $MLP_{nobis}$ and $MLP_{bis}$, configured with the optimized hyperparameters, were fitted using the entire training set.

Figure 4.2 displays the confusion matrices of $MLP_{nobis}$ and $MLP_{bis}$, evaluated in the test set. These matrices show the OSA severity group predicted by both $MLP_{nobis}$ and $MLP_{bis}$ versus the actual OSA severity group from PSG. Using $MLP_{nobis}$, 69.3% of the test patients (52/75) were correctly assigned to their actual group of OSA severity (sum of the main diagonal elements of the confusion matrix). Conversely, $MLP_{bis}$ correctly assigned 76.0% (57/75) of the children to their OSA severity group. Table 4.2 shows diagnostic ability statistics of these models for the AHI severity cutoffs of 5 and 10 e/h. Notice that $MLP_{bis}$ outperformed $MLP_{nobis}$ in terms of Se, Sp, PPV, NPV, LR+, LR−, Acc, and kappa for both cutoffs.

**Table 4.3** DWT-derived feature values for each group (median [interquartile range]) in the optimization set

| Features | Negative OSA | Positive OSA | $p$-value |
|---|---|---|---|
| $M1_{D9}$ | 3.04 [2.26, 3.92] | 5.36 [3.77, 7.70] | <0.01 |
| $M2_{D9}$ | 3.78 [3.23, 4.63] | 5.73 [4.30, 7.57] | <0.01 |
| $M3_{D9}$ | 1.31 [1.20, 1.44] | 1.19 [1.06, 1.32] | <0.01 |
| $M4_{D9}$ ($10^2$) | 3.58 [1.03, 7.69] | 0.06 [0.04, 2.69] | <0.01 |
| $Max_{D9}$ ($10^1$) | 1.23 [1.04, 1.55] | 1.96 [1.42, 2.62] | <0.01 |
| $En_{D9}$ ($10^3$) | 0.54 [0.37, 0.89] | 1.54 [0.78, 2.96] | <0.01 |
| WE ($10^{-4}$) | 1.83 [1.18, 2.86] | 4.27 [2.52, 9.41] | <0.01 |

This table has been derived from Vaquerizo-Villar et al. [2]
OSA = obstructive sleep apnea

### *4.1.2 Wavelet Analysis*

In Vaquerizo-Villar et al. [2], the utility of wavelet analysis to provide discriminative features from overnight oximetry associated to pediatric OSA was examined. To this effect, seven DWT-derived features were extracted for each oximetry signal of a database of 981 children with suspected OSA (i.e. the complete UofC database). In order to assess complimentarity with conventional approaches, ODI3, statistical moments in the time domain, and PSD features were also extracted to obtain a wide initial feature set. FCBF was then employed to select an optimum feature subset. Finally, LR, SVM, and MLP binary classifiers were fitted with the optimum feature subset to detect moderate-to-severe pediatric OSA (AHI $\geq$ 5 e/h). In order to assure a proper validation of the proposed methodology, the database was divided into two sets: optimization set (589 subjects, 60%) and cross-validation set (392 subjects, 40%). The optimization group was used to: (*i*) perform descriptive analysis of the DWT-derived parameters; (*ii*) obtain an optimum subset of features with FCBF; (*iii*) optimize the hyperparameters of the SVM and MLP classifiers. Conversely, the cross-validation group was employed to assess the diagnostic performances of the extracted features and the LR, SVM, and MLP classifiers using stratified 5-fold cross-validation.

Figure 4.3 shows the histogram of the $D_9$ coefficients for the negative OSA (AHI $<$ 5 e/h) and positive OSA (AHI $\geq$ 5 e/h) groups in the optimization set. According to this figure, $D_9$ coefficients are more concentrated near zero in the negative OSA group, whereas in the positive OSA group the coefficients are more disperse. Table 4.3 shows the median and interquartile range of the DWT-derived features in the optimization set for negative OSA and positive OSA groups, along with their corresponding $p$-values, computed according to the Mann-Whitney $U$ test. Noteworthy is the fact that all DWT-derived features showed statistically significant differences ($p$-value $< 0.01$). The values of $M1_{D_9}$, $M2_{D_9}$, $Max_{D_9}$, $En_{D_9}$, and $WE$ were significantly higher in the positive OSA group, whereas $M3_{D_9}$ and $M4_{D_9}$ showed higher values in the negative OSA group.

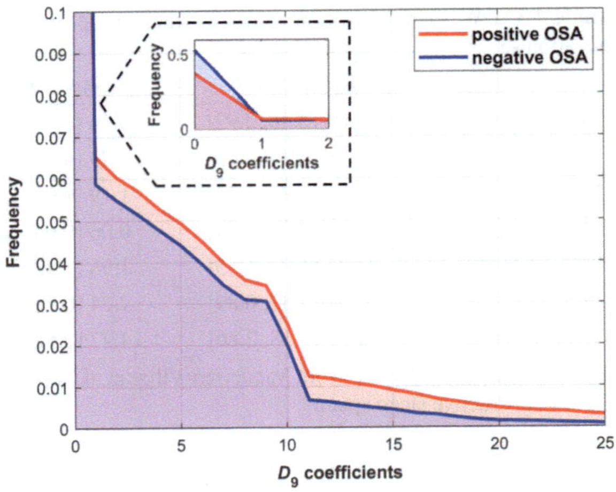


**Fig. 4.3** Histogram of the $D_9$ coefficients for each group in the optimization set. This figure has been derived from Vaquerizo-Villar et al. [2]

Taking as input all the extracted features (ODI3, statistical moments, PSD, and DWT features), FCBF was applied to 1000 bootstrap replicates derived from the optimization set. In this way, $ODI3$, one statistical moment ($M2_t$), three features from PSD ($M2_{BI1}$, $M3_{BI1}$, and $Max_{BI1}$), and three DWT-derived features ($M3_{D_9}$, $En_{D_9}$, and $WE$), which were selected more than 500 times, formed the optimum feature subset. LR, SVM, and MLP binary classifiers were configured using this feature subset. Regarding SVM, we conducted trials varying the regularization parameter ($C$) from $C = 10^{-5}$ to $C = 10^5$ (step of 1 in base-10 logarithmic scale), obtaining $C = 10^3$ as the optimum value for which the accuracy was higher in the optimization set. With respect to MLP, the values of $\alpha$ and $N_H$ were from 0 up to 10 (step of 1) and from 2 up to 50 (step of 1), respectively, obtaining $\alpha = 1$ and $N_H = 5$ as the optimum $N_H - \alpha$ pair that maximized the accuracy in the optimization set.

Table 4.4 displays the diagnostic ability of all the extracted features in the cross-validation set. Notice that 5 out of 7 DWT-derived features reached accuracies near 80%, being $Max_{D_9}$ the DWT-derived feature that achieved the highest performance (81.7±5.6% Acc, with 75.4±7.1% Se and 85.4±6.8% Sp). In this respect, $Max_{D_9}$ showed a similar Acc to ODI3 (81.9±7.2% Acc, with 78.1±7.3% Se and 84.2±8.1% Sp) and outperformed the remaining features. Table 4.5 shows the diagnostic performance metrics of LR, SVM, and MLP models, evaluated in the cross-validation set. Notice that these classifiers showed a high diagnostic ability for the diagnosis of moderate-to-severe OSA (AHI $\geq$ 5 e/h), outperforming all the extracted features individually. Specifically, the SVM binary classifier reached the highest diagnostic performance (84.0±5.2% Acc, with 71.9±4.4% Se and 91.1±7.2% Sp).

**Table 4.4** Diagnostic ability of all the extracted features in the cross-validation set

| Feature | Se | Sp | PPV | NPV | LR+ | LR− | Acc |
|---|---|---|---|---|---|---|---|
| ODI3 | 78.1±7.3 | 84.2±8.1 | 75.2±10.2 | 86.5±5.0 | 6.1±2.9 | 0.27±0.11 | 81.9±7.2 |
| $M1_T$ | 62.3±6.8 | 65.0±2.6 | 51.4±2.1 | 74.6±3.6 | 1.8±0.2 | 0.58±0.10 | 64.0±2.3 |
| $M2_T$ | 72.6±13.6 | 67.1±6.6 | 56.7±2.8 | 81.2±6.6 | 2.2±0.3 | 0.40±0.17 | 69.2±3.1 |
| $M3_T$ | 65.0±8.5 | 61.4±6.8 | 50.1±2.8 | 74.9±2.8 | 1.7±0.2 | 0.57±0.09 | 62.7±2.7 |
| $M4_T$ | 60.9±15.6 | 49.9±8.4 | 41.6±5.0 | 69.0±7.5 | 1.2±0.3 | 0.78±0.26 | 54.0±5.2 |
| $M1_{BI1}$ | 75.3±7.9 | 82.5±7.4 | 73.0±8.5 | 85.1±3.5 | 5.3±3.1 | 0.30±0.08 | 79.9±3.8 |
| $M2_{BI1}$ | 69.8±7.3 | 83.4±5.2 | 71.8±6.2 | 82.5±3.0 | 4.5±1.4 | 0.36±0.08 | 78.3±3.2 |
| $M3_{BI1}$ | 47.2±11.7 | 58.1±11.9 | 40.4±4.1 | 65.0±2.8 | 1.2±0.2 | 0.91±0.12 | 54.1±4.5 |
| $M4_{BI1}$ | 63.6±8.3 | 47.1±6.2 | 41.7±4.2 | 68.7±6.1 | 1.2±0.2 | 0.79±0.23 | 53.3±5.0 |
| $Max_{BI1}$ | 78.1±8.8 | 75.2±9.9 | 66.2±6.9 | 85.6±3.6 | 3.5±1.1 | 0.29±0.09 | 76.3±4.3 |
| $SE_{BI1}$ | 48.6±14.4 | 61.8±11.8 | 43.0±4.8 | 67.3±3.3 | 1.3±0.3 | 0.82±0.12 | 56.9±4.2 |
| $M1_{D9}$ | 73.4±9.1 | 82.6±7.8 | 72.2±10.2 | 84.0±5.1 | 5.2±2.7 | 0.32±0.12 | 79.1±6.2 |
| $M2_{D9}$ | 74.7±6.1 | 81.7±6.5 | 71.5±6.9 | 84.6±3.0 | 4.6±1.7 | 0.31±0.07 | 79.1±3.3 |
| $M3_{D9}$ | 58.3±9.2 | 63.4±6.5 | 48.7±3.1 | 72.1±3.3 | 1.6±0.2 | 0.66±0.10 | 61.5±3.2 |
| $M4_{D9}$ | 71.2±6.7 | 64.6±5.7 | 54.6±3.3 | 79.2±4.0 | 2.0±0.3 | 0.45±0.10 | 67.1±3.5 |
| $Max_{D9}$ | 75.4±7.1 | 85.4±6.8 | 76.0±9.0 | 85.4±4.3 | 6.2±2.8 | 0.29±0.10 | 81.7±5.6 |
| $En_{D9}$ | 78.8±4.4 | 81.7±5.2 | 72.2±5.5 | 86.7±2.4 | 4.6±1.4 | 0.26±0.05 | 80.6±3.4 |
| WE | 76.0±8.2 | 78.4±5.6 | 68.0±3.8 | 84.9±3.5 | 3.6±0.7 | 0.30±0.09 | 77.6±2.5 |

This table has been derived from Vaquerizo-Villar et al. [2]

Se = sensitivity (%), Sp = specificity (%), PPV = positive predictive value (%), NPV = negative predictive value (%), LR+ = positive likelihood ratio, LR− = negative likelihood ratio, Acc = accuracy (%)

**Table 4.5** Diagnostic ability of the LR, SVM, and MLP classifiers in the cross-validation set

| Feature | Se | Sp | PPV | NPV | LR+ | LR− | Acc |
|---|---|---|---|---|---|---|---|
| LR | 72.6±4.7 | 90.2±6.2 | 82.3±8.8 | 84.7±2.8 | 9.8±5.5 | 0.31±0.06 | 83.7±4.9 |
| SVM | 71.9±4.4 | 91.1±7.2 | 83.8±10.8 | 84.5±2.6 | 14.6±12.9 | 0.31±0.06 | 84.0±5.2 |
| MLP | 73.3±6.6 | 89.0±6.9 | 80.7±9.2 | 84.9±3.3 | 9.0±5.8 | 0.30±0.08 | 83.2±5.2 |

This table has been derived from Vaquerizo-Villar et al. [2]

Se = sensitivity (%), Sp = specificity (%), PPV = positive predictive value (%), NPV = negative predictive value (%), LR+ = positive likelihood ratio, LR− = negative likelihood ratio, Acc = accuracy (%)

### *4.1.3 Detrended Fluctuation Analysis*

The usefulness of DFA-derived features to supply further information from the oximetry signal linked to pediatric OSA was explored in Vaquerizo-Villar et al. [3], employing a conventional three-stage feature-engineering methodology. ODI3 and six DFA-derived parameters were first extracted from the $SpO_2$ recording of 981 children (i.e., the complete UofC database). A reduced subset of relevant and non-redundant features was then obtained using the FCBF algorithm. Finally, a MLP

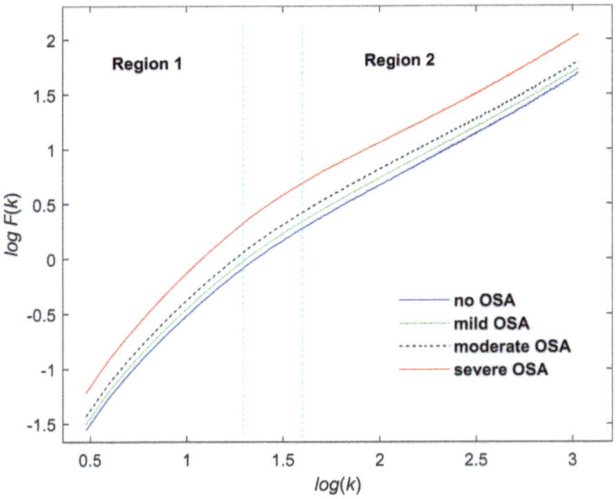


**Fig. 4.4** Averaged DFA profile by OSA severity group in the training dataset. ©2018 Institute of Physics and Engineering in Medicine. Reproduced by permission of IOP Publishing from Vaquerizo-Villar et al. [3]. 

regression neural network was trained and optimized using this feature subset to estimate the AHI and hence pediatric OSA severity degrees. According to this feature-engineering methodology, the database was split into two sets: (*i*) training set (589 subjects, 60%), employed for optimization purposes; and (*ii*) test set (392 subjects, 40%), used to evaluate the diagnostic performance of the developed approach.

Figure 4.4 displays the averaged DFA profile in the training set for the four OSA severity groups: no OSA (AHI <1 e/h), mild OSA ($1 \leq$ AHI $< 5$ e/h), moderate OSA ($5 \leq$ AHI $< 10$ e/h), and severe OSA (AHI $\geq 10$ e/h). In this figure, higher fluctuations are observed as the OSA severity increases. In addition, the two scaling regions (region 1 and region 2) are observed for the four OSA severity groups, as mentioned in Sect. 3.3.1.2. Table 4.6 shows the median and interquartile range values by OSA severity group of each extracted feature, as well as the $p$-value, obtained by means of the Kruskal-Wallis test. ODI3 and 4 out of 6 DFA-derived parameters ($slope_1$, $slope_{12}$, $F(k_{12})$, and $F(k_x)$) showed statistically significant differences ($p$-value $< 0.01$ after Bonferroni correction). Specifically, these features experienced an increasing trend as OSA severity increases.

FCBF was applied to 1000 bootstrap replicates obtained from all the extracted features (ODI3 and DFA-derived parameters) in the training group. The optimum subset was composed of ODI3 and $slope_1$, as these features were selected more than 500 times. Using this feature subset, the MLP network was trained and optimized. Network hyper-parameters, $\alpha$ and $N_H$, were varied from $\alpha = 0$ up to $\alpha = 10$ (step of 1) and from $N_H = 2$ up to $N_H = 30$ (step of 1), obtaining $\alpha = 1$ and $N_H = 5$ as the optimum $N_H - \alpha$ pair that maximized kappa in the training set.

Fig. 4.5 displays the Bland-Altman graphs of ODI3 and the AHI estimated by the MLP network ($AHI_{MLP\text{-}DFA}$) compared to $AHI_{PSG}$. ICC is also shown. The confidence interval was slightly lower in ODI3 than in $AHI_{MLP\text{-}DFA}$ plot (23.3 versus 24.3). However, $AHI_{MLP\text{-}DFA}$ reached a lower absolute mean error than ODI3 (0.75 vs $-1.65$), as well as a higher ICC (0.891 versus 0.866). Figure 4.6 shows the confusion

**Table 4.6** Feature values for the OSA severity groups (median [interquartile range]) in the training set. 

| Features | No OSA | Mild OSA | Moderate OSA | Severe OSA | $p$-value* |
|---|---|---|---|---|---|
| ODI3 | 1.04 [0.52,2.47] | 2.03 [0.93,3.89] | 3.69 [1.94,7.23] | 12.35 [6.65,24.49] | <0.01 |
| $slope_1$ | 1.63 [1.58,1.68] | 1.64 [1.58,1.70] | 1.67 [1.60,1.71] | 1.74 [1.66,1.79] | <0.01 |
| $slope_2$ | 0.96 [0.90,1.05] | 0.95 [0.87,1.03] | 0.92 [0.85,1.02] | 0.94 [0.88,1.01] | 0.18 |
| $slope_{12}$ | 1.66 [1.53,1.82] | 1.69 [1.55,1.87] | 1.77 [1.60,1.94] | 1.82 [1.68,1.95] | <0.01 |
| $k_{12}$ | 1.33 [1.23,1.42] | 1.36 [1.26,1.44] | 1.38 [1.29,1.45] | 1.34 [1.23,1.42] | 0.04 |
| $F(k_{12})$ | 0.01 [-0.18,0.18] | 0.12 [-0.12,0.26] | 0.22 [0.04,0.38] | 0.42 [0.16,0.61] | <0.01 |
| $F(k_x)$ | -0.05 [-0.13,0.04] | 0.02 [-0.07,0.11] | 0.10 [0.00,0.20] | 0.31 [0.18,0.52] | <0.01 |

*$p$-values obtained from the Kruskal-Wallis test after Bonferroni correction, OSA = obstructive sleep apnea

matrices of ODI3 and $AHI_{MLP\text{-}DFA}$ in the test group. Using ODI3, 55.4% of the children (217/392) were rightly asigned to their corresponding OSA severity group, with a kappa of 0.355. By contrast, 60.0% of subjects (235/392) were rightly assigned by $AHI_{MLP\text{-}DFA}$, with a kappa of 0.422. The diagnostic performance metrics of ODI3 and $AHI_{MLP\text{-}DFA}$ for the AHI-based cutoffs of 1, 5, and 10 e/h were derived from these matrices:

- AHI cutoff = 1 e/h. ODI3 obtained 77.0% Acc (83.5% Se and 50.6% Sp) and 0.811 AUC, whereas $AHI_{MLP\text{-}DFA}$ obtained 82.7% Acc (97.1% Se and 23.3% Sp) and 0.813 AUC.
- AHI cutoff = 1 e/h. ODI3 obtained 82.7% Acc (65.1% Se and 93.1% Sp) and 0.883 AUC, whereas $AHI_{MLP\text{-}DFA}$ obtained 81.9% Acc (78.8% Se and 83.7% Sp) and 0.888 AUC.
- AHI cutoff = 10 e/h. ODI3 obtained 89.5% Acc (65.1% Se and 96.1% Sp) and 0.921 AUC, whereas $AHI_{MLP\text{-}DFA}$ obtained 91.1% Acc (77.1% Se and 94.8% Sp) and 0.930 AUC.

Notice that $AHI_{MLP\text{-}DFA}$ outperformed ODI3 in terms of kappa, Acc for the severity cutoffs of 1 and 10 e/h, and AUC for the three cutoffs. In order to provide a more comprehensive comparison between ODI3 and $AHI_{MLP\text{-}DFA}$, ICC, kappa, overall Acc (four classes), and AUC values were obtained from 1000 bootstrap replicates and the

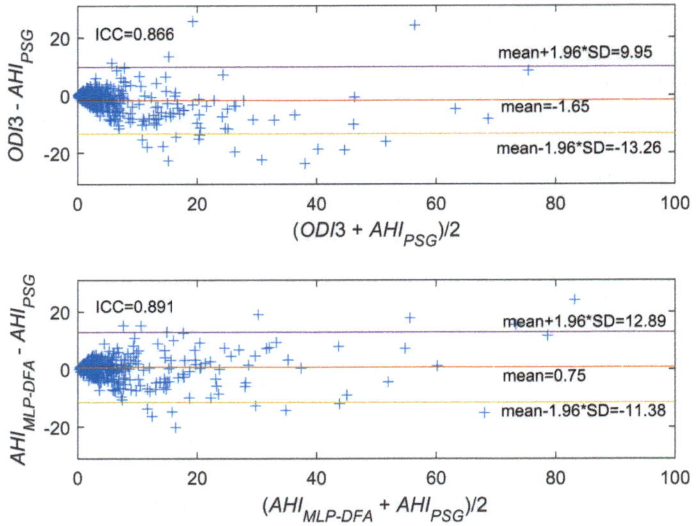


**Fig. 4.5** Bland-Altman plots comparing **a** ODI3 and **b** $AHI_{MLP\text{-}DFA}$ with $AHI_{PSG}$ in the test set. 

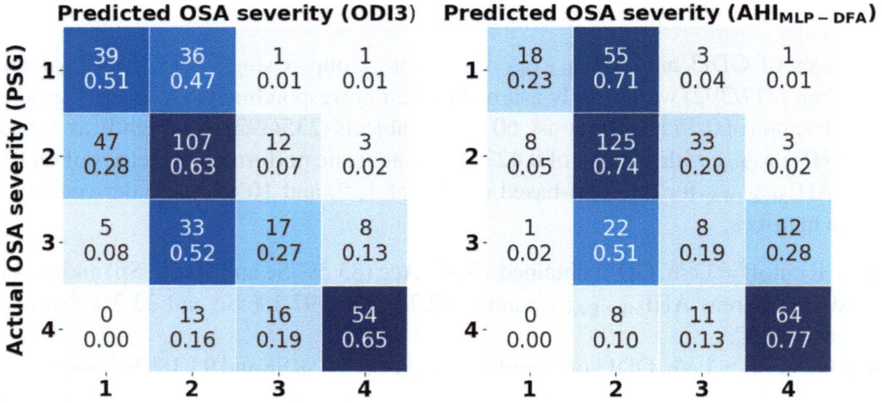


**Fig. 4.6** Confusion matrices of ODI3 and $AHI_{MLP\text{-}DFA}$ in the test set. 1: no OSA; 2: mild OSA; 3: moderate OSA; 4: severe OSA

$p$-value between ODI3 and $AHI_{MLP\text{-}DFA}$ was calculated for each of these performance metrics by means of the Mann-Whitney $U$ test. In this way, statistically significant higher values ($p$-value $< 0.01$) were obtained using $AHI_{MLP\text{-}DFA}$ in the case of ICC, kappa, overall Acc, and AUC for the AHI-based cutoffs of 5 e/h and 10 e/h.

## 4.2 Application of Deep-Learning Techniques

The previously mentioned results (see Sects. 4.1.1, 4.1.2, and 4.1.3) were obtained following a feature-engineering methodology. As stated in Sect. 3.4, this approach may lead to the omission of OSA-related information from the oximetry signal. In Vaquerizo-Villar et al. [4], we evaluate a novel methodology based on deep learning to automatically find the relevant information of the oximetry signal linked to pediatric OSA. This methodology consisted of two steps. First, a CNN architecture was trained to estimate the number of apneic events in each 20-min non-overlapping $SpO_2$ segment. Then, the outputs of the CNN for each segment are aggregated to estimate the AHI for each subject using a database of 3196 $SpO_2$ recordings of children from three independent datasets: the CHAT dataset (see Sect. 3.1.1), the UofC dataset (see Sect. 3.1.2), and the BUH dataset (see Sect. 3.1.3).

In order to assure a proper generalization of the proposed approach, the whole population under study was divided into three sets: (*i*) training set, used to train the CNN architecture; (*ii*) validation set, used for hyperparameter optimization; (*iii*) test set, used to assess the diagnostic performance of the deep-learning model. The training group was composed of 859 $SpO_2$ recordings from the CHAT dataset, as the UofC and BUH databases do not contain respiratory events annotations, which are needed in the CNN architecture to compose the output labels of each 20-min non-overlapping segment from the training set (see Sect. 3.4.1). The remaining subjects from the CHAT dataset, as well as the subjects of the UofC and BUH sets, were randomly divided into a validation set (1402 subjects, 60%) and a test set (40%), composed of 312 children from the CHAT dataset, 392 children from the UofC dataset, and 231 children from the BUH dataset.

Figure 4.7 shows the results of the BO-TPE algorithm for the hyper-parameters of the CNN architecture: the number of filters ($M_C$) and the filter size ($L_C$) in each 1-D convolution, the number of CNN blocks ($\lambda_C$), the delta parameter of the Huber loss function ($\delta$), and the dropout probability ($p_{drop}$). For each hyperparameter, the values of kappa are displayed in a boxplot. It can be observed that there is not a high confidence (interquartile range) of kappa on the values of the hyperparameter. Slightly higher overall kappa values are achieved with an increasing tendency of $M_C$ and a decreasing trend of $\delta$, as well as when $\lambda_C = 6$ and $L_C = 5$, whereas the value of $p_{drop}$ had little effect on kappa. In this way, $M_C = 64$, $L_C = 5$, $\lambda_C = 6$, $\delta = 1.5$, and $p_{drop} = 0.1$ were obtained as the optimum values that maximized kappa in the validation set.

Figure 4.8 shows the scatter plots of the AHI estimated by the CNN model ($AHI_{CNN}$) compared to $AHI_{PSG}$ in the CHAT, UofC, and BUH test sets. ICC and RMSE between $AHI_{CNN}$ and $AHI_{PSG}$ are also shown. It can be observed that points of the scatter plot are more concentrated near the diagonal line in the CHAT test set, which results in a higher agreement (ICC = 0.960 and RMSE = 2.89) than in the UofC (ICC = 0.917 and RMSE = 5.45) and BUH test sets (ICC=0.583 and RMSE = 10.44). Figure 4.9 displays the error distribution plots of $AHI_{CNN}$ in the three test sets. A low mean error was obtained in the three test datasets. However, 95% con-

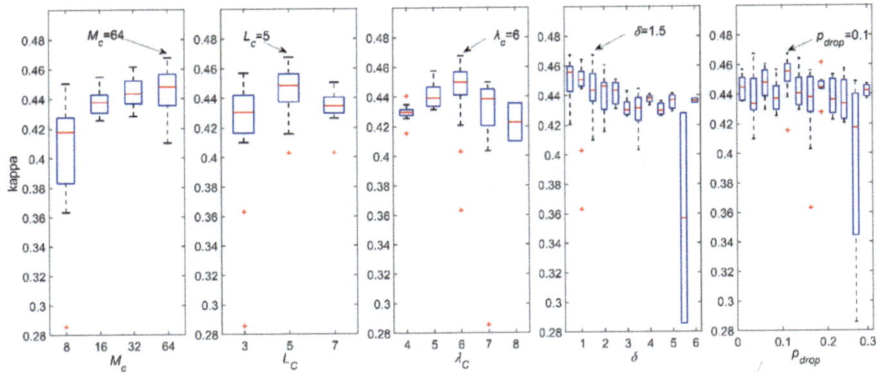


**Fig. 4.7** Results of the hyperparameter optimization in the validation set. ©2021 IEEE. Reprinted, with permission, from Vaquerizo-Villar et al. [4]

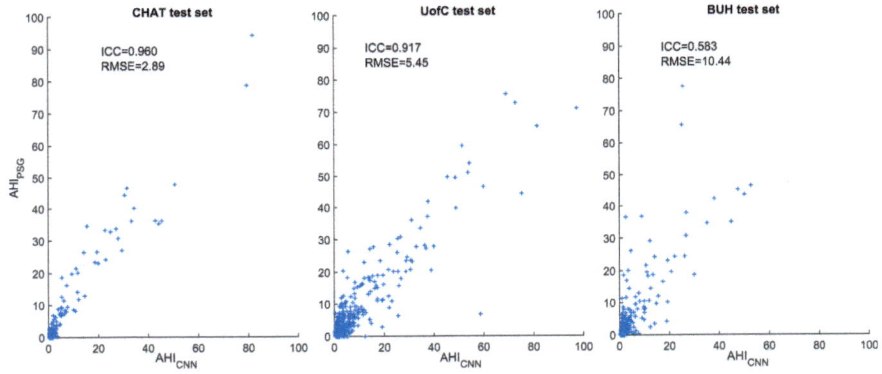


**Fig. 4.8** Scatter plot comparing $AHI_{CNN}$ with $AHI_{PSG}$ in the CHAT, UofC, and BUH test databases. ©2021 IEEE. Reprinted, with permission, from Vaquerizo-Villar et al. [4]

fidence interval of $AHI_{CNN}$ was lower in the CHAT test set (12.80 e/h) than in the UofC (21.69 e/h) and BUH (28.84 e/h) test sets. In this respect, some outliers were observed in the UofC and BUH sets, as indicated by the maximum error.

Figure 4.10 shows the confusion matrices that face the pediatric OSA severity degrees established by the $AHI_{PSG}$ and the corresponding assignation using $AHI_{CNN}$ in the three test sets. Notice that a higher four-class overall accuracy was reached by $AHI_{CNN}$ in the CHAT test set (72.8%, 227/312) than in the UofC (60.2%, 236/392) and BUH test sets (61.0%, 141/231), as anticipated by the scatter and error distribution plots. Table 4.7 shows the diagnostic performance statistics of $AHI_{CNN}$ for each AHI threshold that establishes the pediatric OSA severity degrees (1 e/h, 5 e/h, and 10 e/h), which are derived from the confusion matrix. The value of kappa was remarkably higher in the CHAT test set (0.515) than in the UofC (0.422) and BUH test sets (0.423). A higher diagnostic ability is also observed in the CHAT test set for the AHI-based cutoffs of 5 and 10 e/h.

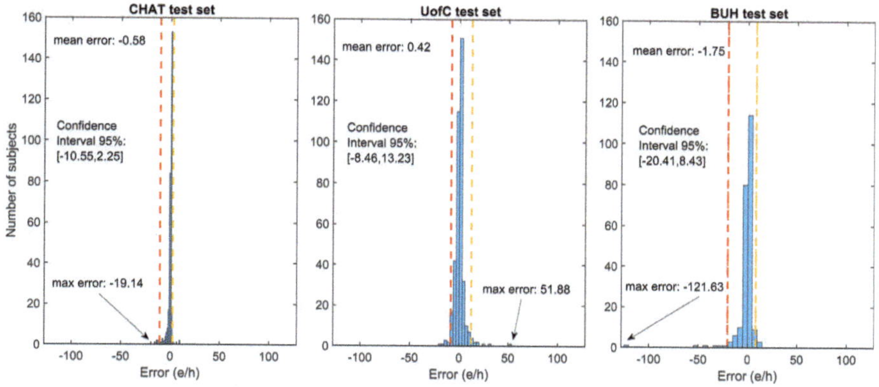


**Fig. 4.9** Error distribution plot of $AHI_{CNN}$ in the CHAT, UofC, and BUH test databases. ©2021 IEEE. Reprinted, with permission, from Vaquerizo-Villar et al. [4]

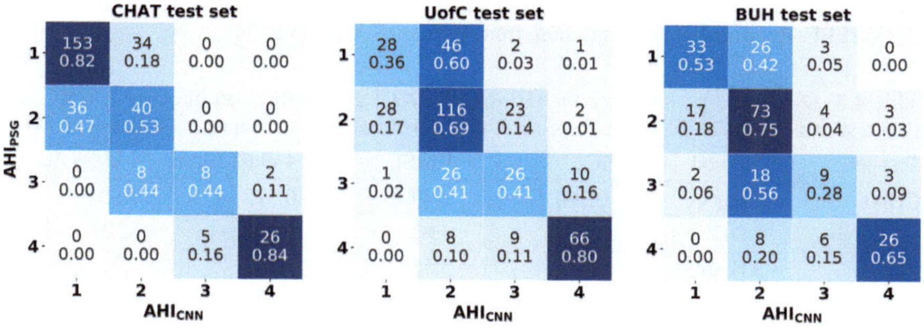


**Fig. 4.10** Confusion matrices of $AHI_{CNN}$ in the CHAT, UofC, and BUH test datasets. 1: no OSA; 2: mild OSA; 3: moderate OSA; 4: severe OSA. ©2021 IEEE. Reprinted, with permission, from Vaquerizo-Villar et al. [4]

In order to provide a thorough comparison between our proposal and conventional approaches, we have compared the results of $AHI_{CNN}$ with ODI3, a clinical parameter, as well as the AHI obtained with a classical feature-engineering approach based on MLP ($AHI_{MLP}$) [4]. Table 4.8 shows the comparison of the performance of $AHI_{CNN}$ with ODI3 and $AHI_{MLP}$ in the three test sets. Notice that $AHI_{CNN}$ outperformed ODI3 and $AHI_{MLP}$ in terms of overall accuracy, kappa, RMSE, and ICC in the three test sets.

In this chapter, the most relevant results obtained during this doctoral thesis have been presented. In the next Chapter (see Sect. 5), these results will be discussed, as well as compared with state-of-the-art studies.

**Table 4.7** Diagnostic ability of $AHI_{CNN}$ for the AHI cutoffs = 1 e/h, 5 e/h, and 10 e/h in the CHAT, UofC, and BUH test databases

| Test set | AHI cutoff | Se | Sp | PPV | NPV | LR+ | LR− | Acc | Kappa |
|---|---|---|---|---|---|---|---|---|---|
| CHAT | **1 e/h** | 71.2 | 81.8 | 72.4 | 81.0 | 3.92 | 0.35 | 77.6 | 0.515 |
| | **5 e/h** | 83.7 | 100 | 100 | 97.0 | N.D. | 0.16 | 97.4 | |
| | **10 e/h** | 83.9 | 99.3 | 92.9 | 98.2 | 117.84 | 0.16 | 97.8 | |
| UofC | **1 e/h** | 90.8 | 36.4 | 85.4 | 49.1 | 1.43 | 0.25 | 80.1 | 0.422 |
| | **5 e/h** | 76.0 | 88.6 | 79.8 | 86.2 | 6.68 | 0.27 | 83.9 | |
| | **10 e/h** | 79.5 | 95.8 | 83.5 | 94.6 | 18.90 | 0.21 | 92.3 | |
| BUH | **1 e/h** | 88.8 | 53.2 | 83.8 | 63.5 | 1.90 | 0.21 | 79.2 | 0.423 |
| | **5 e/h** | 61.1 | 93.7 | 81.5 | 84.2 | 9.72 | 0.42 | 83.5 | |
| | **10 e/h** | 65.0 | 96.9 | 81.3 | 93.0 | 20.69 | 0.36 | 91.3 | |

AHI = apnea-hypopnea index, Se = sensitivity (%), Sp = specificity (%), PPV = positive predictive value (%), NPV = negative predictive value (%), LR+ = positive likelihood ratio, LR− = negative likelihood ratio, Acc = accuracy (%), kappa = Cohen's kappa index, N.D. = not defined


**Table 4.8** Diagnostic performance of $AHI_{CNN}$ vs. ODI3 and $AHI_{MLP}$ in the CHAT, UofC, and BUH test databases. ©2021 IEEE. Reprinted, with permission, from Vaquerizo-Villar et al. [4]

| Test set | Method | ICC | RMSE | 4-class kappa | 4-class Acc |
|---|---|---|---|---|---|
| CHAT | $AHI_{CNN}$ | 0.960 | 2.89 | 0.515 | 72.8 |
| | ODI3 | 0.871 | 4.63 | 0.417 | 65.1 |
| | $AHI_{MLP}$ | 0.832 | 5.51 | 0.377 | 63.3 |
| UofC | $AHI_{CNN}$ | 0.917 | 5.45 | 0.422 | 60.2 |
| | ODI3 | 0.861 | 6.21 | 0.372 | 56.6 |
| | $AHI_{MLP}$ | 0.890 | 6.02 | 0.381 | 56.9 |
| BUH | $AHI_{CNN}$ | 0.583 | 10.44 | 0.423 | 61.0 |
| | ODI3 | 0.520 | 10.64 | 0.369 | 57.6 |
| | $AHI_{MLP}$ | 0.500 | 11.05 | 0.306 | 52.4 |

ICC = intra-class correlation coefficient, RMSE = root mean squared error, kappa = Cohen's kappa index, Acc = accuracy (%)

## References

1. Vaquerizo-Villar F, Álvarez D, Kheirandish-Gozal L, Gutiérrez-Tobal GC, Barroso-García V, Crespo A, del Campo F, Gozal D, Hornero R (2018) Utility of bispectrum in the screening of pediatric sleep apnea-hypopnea syndrome using oximetry recordings. Comput Meth Programs Biomed 156:141–149
2. Vaquerizo-Villar F, Álvarez D, Kheirandish-Gozal L, Gutiérrez-Tobal GC, Barroso-García V, Crespo A, del Campo F, Gozal D, Hornero R (2018) Wavelet analysis of oximetry recordings to assist in the automated detection of moderate-to-severe pediatric sleep apnea-hypopnea syndrome. PloS one 13(12):e0208502
3. Vaquerizo-Villar F, Álvarez D, Kheirandish-Gozal L, Gutiérrez-Tobal GC, Barroso-García V, Crespo A, Del Campo F, Gozal D, Hornero R (2018) Detrended fluctuation analysis of the

oximetry signal to assist in paediatric sleep apnoea-hypopnoea syndrome diagnosis. Physiol Meas 39(11):114006

4. Vaquerizo-Villar F, Alvarez D, Kheirandish-Gozal L, Gutierrez-Tobal GC, Barroso-Garcia V, Santamaria-Vazquez E, Del Campo F, Gozal D, Hornero R (2021) A convolutional neural network architecture to enhance oximetry ability to diagnose pediatric obstructive sleep apnea. IEEE J Biomed Health Inform 25(8):2906–2916

# Chapter 5
# Discussion

This doctoral thesis addresses the simplification of pediatric OSA diagnosis. For this purpose, novel signal processing algorithms have been applied to improve the diagnostic ability of the oximetry signal. In this regard, feature-engineering and deep-learning methodologies have been approached. On the one hand, we have gone further in $SpO_2$ characterization in the time and frequency domain using bispectrum, wavelet, and DFA, obtaining new features that provide additional information from the oximetry signal related to pediatric OSA and its severity. On the other hand, a deep-learning model based on CNNs was able to automatically learn discriminative features from raw $SpO_2$ data linked to apneic events, outperforming conventional approaches. In this chapter, the main outcomes obtained during this thesis are discussed. Moreover, a comparison of the proposed methodologies in terms of diagnostic performance is provided, as well as a comparison with state-of-the-art works. Finally, the main limitations of this thesis are stated.

## 5.1 Novel Features to Provide Relevant and Complementary Information from Oximetry Recordings

As aforementioned, bispectrum [1], wavelet analysis [2], and DFA [3] were applied to identify features able to provide additional information regarding OSA-related changes in the oximetry dynamics.



F. Vaquerizo Villar, *Automated Analysis of the Oximetry Signal to Simplify the Diagnosis of Pediatric Sleep Apnea*, Springer Theses,
https://doi.org/10.1007/978-3-031-32832-9_5

### 5.1.1 Bispectral Analysis

Spectral analysis has been widely used to analyze $SpO_2$ in both adult and pediatric OSA contexts [4–8], as it reflects the changes in the $SpO_2$ spectrum elicited by the recurrence of respiratory events while sleeping. This analysis has been commonly accomplished using PSD [5–8]. However, PSD cannot characterize changes of linearity and gaussianity in a signal, as the phase relationship among spectral components is lost [9]. In contrast, bispectral analysis preserves both amplitude and phase information of the Fourier transform, which enables the detection of phase relationships and deviations from gaussianity and linearity of a signal [9], such as those that may be elicited in $SpO_2$ recordings by physiological perturbations of OSA.

In the present doctoral thesis bispectrum has been applied as a complementary tool to conventional spectral analysis [1]. To our knowledge, this is the first time that bispectral analysis is used in the framework of pediatric OSA. A MLP model fed with a feature subset composed of bispectrum-derived parameters, together with anthropometric variables, ODI3, and PSD-derived parameters ($MLP_{bis}$) reached a high diagnostic performance, with a 3-class Acc of 76%, a kappa value of 0.56, and 81.3% Acc and 85.3% Acc for the AHI cutoffs of 5 e/h and 10 e/h, respectively. These results outperformed a MLP neural network trained without information from bispectrum ($MLP_{nobis}$). It is worthy to note that $MLP_{bis}$ reached a PPV of 95.5% for 5 e/h and a NPV of 86.7% for 10 e/h. As stated in Sect. 3, these cutoffs are commonly use in the clinical practice to detect moderate ($5 \leq AHI < 10$ e/h) and severe OSA ($AHI \geq 10$ e/h). In this respect, adenotonsillectomy treatment is recommended in children with an $AHI \geq 5$ e/h, as they have an increased chance of suffering adverse health consequences and comorbidities [10–12]. Furthermore, children with an AHI $\geq 10$ e/h can present persistent risk factors and residual OSA after treatment [13].

It is also important to highlight that two bispectral-derived features, $M1_{BISP}$ and $meanPa_{BISP}$, were involved in the optimum subset. $M1_{BISP}$ contains information about changes in the amplitude of the bispectrum related to deviations of gaussianity in the $SpO_2$ signal, whereas $meanPa_{BISP}$ measures changes in the bispectral phase associated to a phase coupling between spectral components of the $SpO_2$ signal related to pediatric OSA severity. Furthermore, bispectral moments ($H1_{BISP}$, $H2_{BISP}$, and $H3_{BISP}$), which measure nonlinear relationships between the frequency components of the oximetry signal, showed significantly higher values as the severity of OSA increased.

These findings agree with previous works that also examined the usefulness of bispectrum to characterize OSA-related changes in adults [14, 15] and children [16]. Tagluk and Sezgin [15] reported changes in the quadratic phase coupling of the EEG signal, whereas Atri and Mohebbi [14] reported changes in the non-Gaussian and nonlinear dynamical information of heart rate variability and ECG-derived respiratory signals during apneic episodes by means of bispectrum. Finally, in a recent work developed by Barroso-García et al. [16], it was found that bispectrum provides information regarding changes in the gaussianity, linearity, and regularity of the AF signal elicited by apneic events. In the present research, it has been demonstrated that bispectral analysis can identify phase relationships and deviations from linearity and gaussianity of the $SpO_2$ signal that provide additional and complementary information to conventional approaches in the automated detection of childhood OSA. This study was the starting point for the second study of the doctoral thesis in which we apply the wavelet transform to further characterize the $SpO_2$ signal in the frequency domain.

### 5.1.2 Wavelet Analysis

PSD and bispectrum are frequency domain analysis techniques based on the STFT [9]. STFT uses a fixed length window to analyze each segment of the signal, assumming that it is stationary [17]. Nonetheless, non-stationary changes occur in the oximetry signal during sleep, mainly due to apneic events [18]. This limitation is overcome by the WT, which does not make assumptions about the stationarity of the signal [17]. Wavelet analysis provides an optimal time-frequency resolution (high frequency resolution at low frequencies and high temporal resolution at high frequencies) [17], which is useful to analyze OSA-related information at the low frequency components of the $SpO_2$ recordings.

In this research, wavelet analysis has been employed to further characterize the oximetry dynamics related to the presence of moderate-to-severe pediatric OSA [2]. Previous studies have shown the usefulness of the WT to characterize the changes in physiological signals related to apneic episodes in adult patients [15, 19–23], though it had not been previously applied in the screening of childhood OSA. Our results revealed that all the DWT-derived features, $WE$ and features computed from the coefficients in $D_9$ ($M1_{D_9}$, $M2_{D_9}$, $M3_{D_9}$, $M4_{D_9}$, $Max_{D_9}$, and $En_{D_9}$), showed statistically significant differences between negative OSA (AHI $< 5$ e/h) and positive OSA (AHI $\geq 5$ e/h) groups. Furthermore, these features reached a overall higher performance than statistical moments in the time domain and PSD-derived parameters, thus suggesting that DWT is a suitable tool to identify OSA-related changes occurring in the oximetry signal. Finally, MLP, LR, and SVM binary classifiers fed with an optimum subset composed of features from these complementary approaches (DWT, statistical moments, ODI3, and PSD) reached a high diagnostic performance, improving the diagnostic ability of all the extracted features. Noteworthy, the SVM model reached the highest Acc (84.0%), Sp (91.1%), PPV (83.8%), and LR+ (14.6%) among the individual features and binary classification algorithms. A high LR+ (LR$>$10) is considered to present solid evidence to determine the presence of a disease, which indicates that this model is especially useful as a screening method to confirm the presence of moderate-to-severe pediatric OSA (AHI $\geq 5$ e/h). Accordingly, our DWT-based SVM model could be used to automatically detect moderate-to-severe pediatric OSA at patient's home, thus reducing associated healthcare costs and intrusiveness of overnight PSG.

Importantly, three DWT-derived parameters, $M3_{D_9}$, $En_{D_9}$, and $WE$, were automatically selected with FCBF. As shown, $M3_{D_9}$ was significantly lower in the positive-OSA group, which indicates that apneas and hypopneas modify the frequency distribution of $SpO_2$ signal and increase its frequency components in the $D_9$ band (0.0244–0.0488 Hz), thus resulting in values less proximal to zero. Regarding $En_{D_9}$, it was higher in the positive-OSA group, which agrees with a higher occurrence of respiratory events that increase the amplitude of the $D_9$ coefficients. These changes of the $SpO_2$ signal in the $D_9$ band are linked to the recurrence and duration of the oxygen desaturation associated to apneic episodes. In addition, $WE$ revealed a higher irregularity in the positive-OSA group, which indicates that pediatric OSA disturbs the energy distribution of the DWT decomposition of the $SpO_2$ signal. According to our results, the information about the occurrence of apneic events provided by DWT through the amplitude ($En_{D_9}$) and the concentration of the $D_9$ coefficients near zero ($M3_{D_9}$), and the irregularity of the energy distribution ($WE$) of the $SpO_2$ is complementary (non-redundant) to the information provided by conventional approaches. Hence, this study confirms that the great resolution provided by DWT in the low frequency range, as well as its suitability to analyze non-stationary signals, make DWT an appropriate tool to further characterize the changes occurring in the oximetry signal associated with pediatric OSA. These findings, together with those obtained in the Sect. 5.1.1, led us to the third study of this thesis in which we apply the detrended fluctuation analysis method to gain insight into the nonlinear and non-stationary properties of the oximetry signal in the time domain.

### *5.1.3 Detrended Fluctuation Analysis*

As aforementioned, biomedical signals typically present non-stationarities and nonlinearities, since biological systems have an stochastic behavior. In this respect, nonlinear methods derived from the chaos theory have proved high capability to characterize changes in $SpO_2$ dynamics related to physiological perturbations of OSA, both in adult and in pediatric patients [24]. Nonetheless, Garde et al. [7] and Hornero et al. [8] reported that conventional nonlinear metrics (SampEn, CTM, and LZC) were redundant with regard to common statistical moments, conventional oximetric indices, and frequency domain features. In order to provide further insights into its nonlinear properties, we have applied DFA to characterize changes in the scaling behavior (i.e., irregular fluctuations and random spikes) of the oximetry signal related to pediatric OSA and its severity [3]. Previous studies have assessed the capability of DFA to characterize OSA in adults [25–28] and children [29]. Nonetheless, no studies have applied DFA to analyze $SpO_2$ recordings in the context of childhood OSA.

Our results revealed that the scaling behavior of the $SpO_2$ recordings is affected by pediatric OSA. This agrees with Penzel et al. [28] and Dehkordi et al. [29], who also obtained two scaling regions with different correlation properties in adult OSA patients, one region for short-time scales related to respiratory events and another region for long-time scales associated to the effects of circadian rhythm and sleep stages. As shown, $slope_1$ and $slope_{12}$ showed significantly higher values as the AHI increased. This can be explained by the higher occurrence of respiratory events that changes the oximetry dynamics in the short-time scales. In addition, $F(k_{12})$ and $F(k_x)$ showed significantly higher values as the OSA severity increased. This fact is consistent with the higher fluctuations observed in the DFA profile as AHI increased, which indicates that apneic events cause irregular fluctuations in the oximetry signal, as also reported by Hua and Yu [25].

The information provided by DFA was also complimentary to ODI3. Specifically, FCBF automatically selected $slope_1$, together with ODI3. As shown, a MLP neural network trained to estimate the AHI using this optimum subset reached an ICC of 0.891, a 4-class Acc of 60%, a kappa value of 0.41, and 82.7%, 81.9%, and 91.1% Acc for the AHI cutoffs of 1 e/h, 5 e/h, and 10 e/h, respectively. This MLP model showed an overall higher diagnostic performance than the conventional ODI3. These results suggest that the changes in the scaling behavior of the DFA profile quantified by $slope_1$ provide relevant and additional information that contributes to improve the diagnostic ability of the $SpO_2$ signal in the framework of pediatric OSA.

## 5.2 A Deep-Learning Based Methodology to Automatically Extract the Relevant Information from Raw Oximetry Recordings

The feature-engineering approach has shown its usefulness to characterize pediatric OSA and its severity [1–3]. It has been demonstrated that the application of novel signal processing algorithms from different analytical approaches provide relevant features that parameterize OSA-related oximetric changes. Furthermore, this feature-engineering approach has allowed us to identify which features provide additional information to classical methods regarding oximetric changes related to childhood OSA and its severity [1–3]. Nonetheless, the feature-engineering approach is limited to the existing human knowledge, which may lead to the omission of relevant information concerning pediatric OSA that is still undiscovered [30].

The aforementioned limitation of the feature-engineering methodology is overcome by the deep-learning approach, which are based on an automatic identification of the important information that is not controlled by human experts [31]. In Vaquerizo-Villar et al. [32], we have proven that a CNN-based deep-learning method can automatically learn discriminative information from the raw oximetry data linked to apneic events. These findings are consistent with recent studies that also demonstrated the utility of deep-learning approaches to automatically identify OSA-related changes in physiological signals from adult subjects [33–36]. These studies analyzed raw signals from PSG using deep-learning models based on recurrent neural networks (RNN) [33, 36], MLP [35], and CNN [34]. In this regard, CNN has a lower computational cost than RNN and MLP, which makes it more suitable for screening purposes using wearable and portable pulse oximetry devices.

Table 5.1 shows a summary of the comparison of the estimated AHI from the proposed CNN architecture ($AHI_{CNN}$) with ODI3, a clinical approach, as well as the AHI estimated by a classical feature-engineering approach ($AHI_{MLP}$). As expected,

**Table 5.1** Summary of the comparison of $AHI_{CNN}$ with ODI3 and $AHI_{MLP}$ in the CHAT, UofC, and BUH test databases

| Test set | Method | kappa |
|---|---|---|
| CHAT | $AHI_{CNN}$ | 0.515 |
| | ODI3 | 0.417 |
| | $AHI_{MLP}$ | 0.377 |
| UofC | $AHI_{CNN}$ | 0.422 |
| | ODI3 | 0.372 |
| | $AHI_{MLP}$ | 0.381 |
| BUH | $AHI_{CNN}$ | 0.423 |
| | **ODI3** | 0.369 |
| | **$AHI_{MLP}$** | 0.306 |

*kappa* = Cohen's kappa index

$AHI_{CNN}$ outperformed ODI3 and $AHI_{MLP}$, showing a high diagnostic ability in a large sample of 3196 $SpO_2$ signals from three independent datasets. Specifically, $AHI_{CNN}$ reached a high agreement in the CHAT (ICC = 0.960), UofC (ICC = 0.917), and BUH (ICC = 0.583) test sets. In addition, high 4-class accuracies (72.8%, 60.2%, and 61.0%), high kappa values (0.515, 0.422, and 0.423), and high accuracies for the AHI severity cutoffs of 1 e/h (77.6%, 80.1%, and 79.2%), 5 e/h (97.4%, 83.9%, and 83.5%), and 10 e/h (97.8%, 92.3%, and 91.3%) were obtained in the CHAT, UofC, and BUH test sets, respectively.

Fig. 5.1 displays a possible screening protocol that shows the clinical applicability of $AHI_{CNN}$. This screening protocol, which is derived from the confusion matrices of $AHI_{CNN}$ (see Fig. 4.10), would act as follows: (*i*) If $AHI_{CNN}$ <1 e/h (no OSA), clinicians could discard OSA as 96.2% (BUH), 98.2% (UofC), and 100% (CHAT) of these subjects will have an $AHI_{CNN}$ < 5 e/h. These patients might be eventually referred to PSG on the persistence of symptoms [13]; (*ii*) If $1 \leq AHI_{CNN} < 5$ e/h, the clinicians could suggest to conduct overnight PSG as doubts exist about the true diagnosis of these subjects; (*iii*) If $5 \leq AHI_{CNN} < 10$ e/h, the clinicians could consider treatment as 86.4% (BUH), 96.7% (UofC), and 100% (CHAT) of these children are at least mild OSA ($AHI_{CNN} \geq 1$ e/h) and they were referred to the sleep laboratory showing symptoms; (*iv*) If $AHI_{CNN} \geq 10$ e/h, the clinicians could suggest treatment, as most probably (90.6% in BUH, 96.2% in UofC, and 100% in CHAT) these children are at least moderate OSA ($AHI_{CNN} \geq 5$ e/h), and also consider a follow-up of these children, as they are prone to present persistent risk factors and residual OSA after being treated. The implementation of this screening protocol in a pediatric sleep unit could lead to a 45.9% (BUH), 50.0% (UofC), and 73.7% (CHAT) reduction in full PSGs, thus reducing health costs and waiting lists. In addition, these children would benefit from a more comfortable diagnostic test that could be easily performed at home.

Despite the fact that the CNN model outperformed conventional methods in the three datasets, it is noteworthy that our proposal reached a higher performance in the

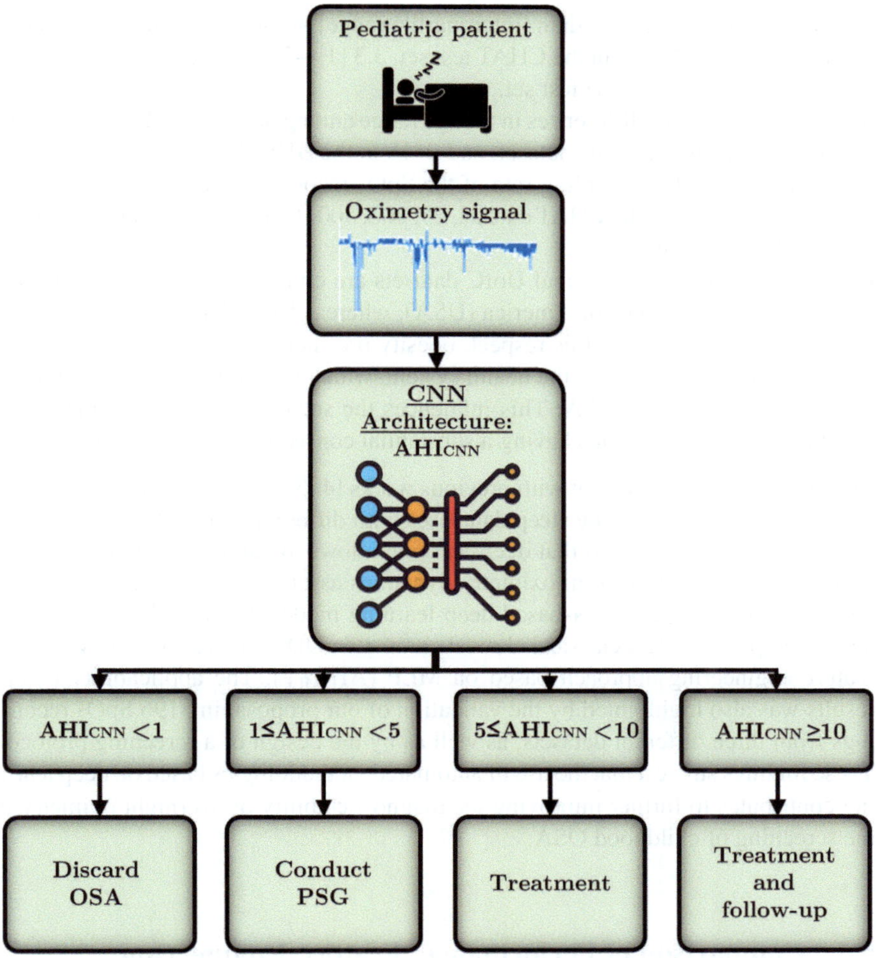


**Fig. 5.1** Screening protocol of the proposed CNN-based deep-learning model

CHAT dataset than in the UofC and BUH datasets. This is consistent with the fact that the weights of the CNN model were obtained using only the CHAT dataset. As there is a large variability in the annotation of cardiorespiratory events and sleep stages among different sleep technologists [37], this variance may influence the external assessment of our proposed CNN model in two external datasets. Nonetheless, we tried to reduce this variability by using a validation group formed by children from the CHAT, UofC, and BUH datasets. There are also differences in the following clinical characteristics that could help to explain the varying diagnostic performance among datasets:

- AHI distribution. The median [interquartile range] of the AHI values vary among datasets: 0.8 [0.4–1.7] in the CHAT test set, 3.3 [1.4–7.8] in the UofC test set, and 2.3 [0.9–6.4] in the BUH test set.
- Age. There are also differences in the age range among datasets: 5–10 in the CHAT dataset, 0–13 in the UofC dataset, and 0–18 in the BUH dataset.
- Sampling rate. The sampling rate of the $SpO_2$ recordings were: 1, 2, 10, 12, 16, 200, 256, 512 Hz in the CHAT dataset, 25, 200, 500 Hz in the UofC dataset, 200 Hz in the BUH dataset.
- Population group. CHAT and UofC datasets are composed of pediatric subjects from the United States of America (USA), whereas BUH dataset is composed of children from Spain. In this respect, obesity prevalence and race are different in these countries, as well as the health system, which is mainly public in Spain and mostly private in the USA. This influences the socioeconomic condition of the subjects under study, thus having a substantial consequence on the health status.

These findings are consistent with previous works [4, 33, 35], which also reported a varying performance among sleep databases with different clinical features.

To sum up, it was found that deep-learning allows to automatically identify discriminative information from oximetry dynamics associated to apneic episodes. In addition, the proposed CNN-based deep-learning model showed a high diagnostic ability, outperforming a classical clinical parameter, ODI3, as well as a conventional feature-engineering approach based on MLP ($AHI_{MLP}$). The applicability of our results was also highlighted by the validation of our proposal in 3196 $SpO_2$ recordings from three different datasets, as well as by the design of a screening protocol. These findings suggest that the use of automated methodologies based on deep learning contributes to further improving the diagnostic ability of overnight oximetry in the screening of childhood OSA.

## 5.3 Comparison of Performance: Feature-Engineering, Deep-Learning, and State-of-the-art

In order to further discuss the most relevant findings, we have compared the diagnostic performance obtained with the different feature-engineering and deep-learning methodologies proposed in the papers included in this thesis (Sect. 5.3.1). In addition, we have compared these results with those achieved in the state-of-the-art, considering the methodological differences among studies that limit generalization. (Section 5.3.2).

### 5.3.1 Comparison Between Feature-Engineering and Deep-Learning Approaches

Table 5.2 displays the overall diagnostic performance of the automated feature-engineering and deep-learning models obtained in the publications of the thesis. Notice that a high performance was achieved in all the publications, with accuracies ranging between 77.6 and 82.7% Acc for an AHI cutoff of 1 e/h, 81.3–97.4% Acc using an AHI cutoff of 5 e/h, and 85.3–97.8% using an AHI cutoff of 10 e/h. It can also be observed that, in Vaquerizo-Villar et al. [1], the performance was lower than in the remaining studies. Nonetheless, an initial version of the UofC database (298 children) was used in this study, so that a thorough comparison of the results cannot be performed. Conversely, the test set of the UofC dataset employed in Vaquerizo-Villar et al. [2], Vaquerizo-Villar et al. [3], and Vaquerizo-Villar et al. [32] was composed of the same subjects (392 pediatric patients). This enables a direct comparison of their results.

Regarding this comparison, it is important to note that a high diagnostic performance was obtained in Vaquerizo-Villar et al. [32] in the UofC dataset. The proposed CNN-based deep-learning model outperformed the DFA-based feature-engineering approach designed in Vaquerizo-Villar et al. [3], achieving a slightly higher 4-class kappa and ICC, as well as higher accuracies for the AHI cutoffs of 5 and 10 e/h than Vaquerizo-Villar et al. [3]. This superior performance is even more noteworthy considering that the optimum CNN model obtained in Vaquerizo-Villar et al. [32] was trained using only the CHAT dataset, whereas the DFA-based MLP model in Vaquerizo-Villar et al. [3] was trained and optimized in the UofC dataset. This highlights the generalization ability of the proposed deep-learning model. Additionally, a similar performance was obtained in Vaquerizo-Villar et al. [32] and Vaquerizo-Villar et al. [2] for the AHI cutoff of 5 e/h. Apart from the aforementioned differences in the training set, in Vaquerizo-Villar et al. [2] we only focused on binary classification for the AHI cutoff of 5 e/h, whereas in Vaquerizo-Villar et al. [32] we assessed an AHI estimation model, thus emphasizing the reliability of the deep-learning solution.

Despite the fact that a comprehensive comparison with the remaining studies is not possible, in Vaquerizo-Villar et al. [32] we also achieved a high performance in the CHAT and BUH datasets, specially in the CHAT dataset, where outstanding values of kappa (0.52), ICC (0.960), and accuracies (higher than 95%) for the AHI cutoffs of 5 and 10 e/h were obtained. This reinforces the suitability of deep-learning approaches to identify OSA-related hidden patterns from the oximetry signal in a pediatric OSA context. In addition, our CNN-based model is fed with raw data, thus not requiring human-driven knowledge regarding the $SpO_2$ information needed. Nonetheless, the interpretation and explanation of the features learned by the CNN is more difficult.

### 5.3.2 Comparison with State-of-the-art Studies

Tables 5.3 and 5.4 show the details of state-of-the-art studies aimed at simplifying childhood OSA diagnosis by the use of the oximetry signal. Table 5.3 summarizes

**Table 5.2** Summary of the diagnostic performance of the methods developed during the publications of the doctoral thesis

| Study | Dataset | N (Total/test) | AHI | Methods (Feature/classification) | Se | Sp | Acc | kappa | ICC |
|---|---|---|---|---|---|---|---|---|---|
| Vaquerizo-Villar et al. [1] | UofC | 298/75 | 5 | Bispectrum, PSD features, ODI3, and anthropometric variables / Multiclass MLP | 61.8 | 97.6 | 81.3 | 0.56 | – |
| | | | 10 | | 60.0 | 94.5 | 85.3 | | |
| Vaquerizo-Villar et al. [2] | UofC | 981/392 | 5 | ODI3, statistical moments, PSD, and DWT features / Binary SVM | 71.9 | 91.1 | 84.0 | – | – |
| Vaquerizo-Villar et al. [3] | UofC | 981/392 | 1 | DFA and ODI3 / Regression MLP | 97.1 | 23.3 | 82.7 | 0.41 | 0.891 |
| | | | 5 | | 78.8 | 83.7 | 81.9 | | |
| | | | 10 | | 77.1 | 94.8 | 91.1 | | |
| Vaquerizo-Villar et al. [32] | UofC | 3196/392 | 1 | CNN architecture | 90.8 | 36.4 | 80.1 | 0.42 | 0.917 |
| | | | 5 | | 76.0 | 88.1 | 83.9 | | |
| | | | 10 | | 79.5 | 95.8 | 92.3 | | |
| Vaquerizo-Villar et al. [32] | CHAT | 3196/312 | 1 | CNN architecture | 71.2 | 81.8 | 77.6 | 0.52 | 0.960 |
| | | | 5 | | 83.7 | 100 | 97.4 | | |
| | | | 10 | | 83.9 | 99.3 | 97.8 | | |
| Vaquerizo-Villar et al. [32] | BUH | 3196/231 | 1 | CNN architecture | 88.8 | 53.2 | 79.2 | 0.42 | 0.583 |
| | | | 5 | | 61.1 | 93.7 | 83.5 | | |
| | | | 10 | | 65.0 | 96.9 | 91.3 | | |

N = Number of subjects, AHI = apnea-hypopnea index, ICC = intra-class correlation coefficient, kappa = Cohen's kappa index, Se = sensitivity (%), Sp = specificity (%), Acc = accuracy (%)

**Table 5.3** Summary of state-of-the-art studies based on conventional oximetric indices

| Study | N (Total/test) | AHI | Methods (Feature/classification) | Validation | Se | Sp | Acc |
|---|---|---|---|---|---|---|---|
| Kirk et al. [41] | 58/58 | 5 | ODI3 / Thresholding | Direct validation[a] | 66.7 | 60.0 | 64.0 |
| Tsai et al. [43] | 148/148 | 1 | ODI4 / Thresholding | No | 77.7 | 88.9 | 79.0 |
| | | 5 | | | 83.8 | 86.5 | 85.1 |
| | | 10 | | | 89.1 | 86.0 | 87.1[b] |
| Brouillete et al. [38] | 349/349 | 1 | Clusters of desaturations / Thresholding | Direct validation[a] | 42.9 | 97.8 | 64.7 |
| Velasco et al. [45] | 167/167 | 1 | Clusters of desaturations / Thresholding | Direct validation[a] | 86.6 | 98.9 | 93.4[b] |
| Van Eyck et al. [44] | 130/130 | 2 | Brouillete et al. [38] criteria and Velasco et al. [45] criteria | Direct validation[a] | 58 | 88 | 78 |
| | | 5 | | | 66 | 69 | 68 |
| Chang et al. [39] | 141/141 | 5 | ODI3 and sympthoms / Binary LR | Direct validation[a] | 60.0 | 86.0 | 71.6 |
| Villa et al. [46] | 268/268 | 1 | Clusters of desaturations and clinical history / Thresholding | Direct validationa | 91.6 | 40.6 | 85.8 |
| | | 5 | | | 40.6[b] | 97.9[b] | 69.4[b] |
| Ma et al. [42] | 32/32 | 1 | ODI4 / Thresholding | No | 59.3 | 80.0 | 62.5 |
| | | 5 | | | 70.6 | 66.7 | 68.8 |
| | | 10 | | | 64.3 | 83.3 | 75.0[b] |

[a] Direct validation of a scoring criteria against AHI from PSG, [b] Computed from reported data
N = Number of subjects, AHI = apnea-hypopnea index, Se = sensitivity (%), Sp = specificity (%), Acc = accuracy (%)

**Table 5.4** Summary of state-of-the-art studies based on automated signal processing approaches

| Study | N (Total/test) | AHI | Methods (Feature / classification) | Validation | Se | Sp | Acc |
|---|---|---|---|---|---|---|---|
| Garde et al. [7] | 146/146 | 5 | Classical indices, statistical moments, PSD, and nonlinear features / Binary LDA | Four-fold cross validation | 80.0 | 83.9 | 78.5 |
| Álvarez et al. [5] | 50/50 | 1 | Classical indices, statistical moments, PSD, and nonlinear features / Binary LR | Bootstrapping | 89.6 | 71.5 | 85.5 |
| | | 3 | | | 82.9 | 84.4 | 83.4 |
| | | 5 | | | 82.2 | 83.6 | 82.8 |
| Crespo et al. [48] | 146/146 | 3 | Classical indices and nonlinear features / Binary LR | Bootstrapping | 84.5 | 83.0 | 83.5 |
| Hornero et al. [8] | 4191/3602 | 1 | ODI3, statistical moments, PSD, and nonlinear features / Regression MLP | Training-Test | 84.0 | 53.2 | 75.2 |
| | | 5 | | | 68.2 | 87.2 | 81.7 |
| | | 10 | | | 68.7 | 94.1 | 90.2 |
| Crespo et al. [6] | 176/176 | 1 | Classical indices, statistical moments, PSD, and nonlinear features / LR | Bootstrapping | 93.9 | 37.8 | 84.3 |
| | | 3 | | | 74.6 | 81.7 | 77.7 |
| | | 5 | | | 70.0 | 91.4 | 82.7 |
| Xu et al. [49] | 432/432 | 1 | ODI3, statistical moments, PSD, and nonlinear features / Regression MLP | Training-Test | 95.3 | 19.1 | 79.6 |
| | | 5 | | | 77.8 | 80.5 | 79.4 |
| | | 10 | | | 73.5 | 92.7 | 88.2 |
| Álvarez et al.[47] | 142/142 | 5 | Classical indices, statistical moments, PSD, nonlinear, and anthropometric features / Binary LR | Bootstrapping | 73.5 | 89.5 | 83.3 |

N = Number of subjects, AHI = apnea-hypopnea index, Se = sensitivity (%), Sp = specificity (%), Acc = accuracy (%)

the results reported in previous studies based on conventional oximetric indices, whereas Table 5.4 displays results achieved by recent studies using automated signal processing approaches.

As shown in Table 5.3, the diagnostic ability of ODI and clusters of desaturations has been widely assessed [38–46], including in some cases common symptoms [39] and clinical history [46]. Regarding their diagnostic performance, these studies achieved accuracies ranging 62–93% using the AHI threshold of 1 e/h [38, 42, 43, 45, 46], 64–85% for an AHI threshold of 5 e/h [39, 41–43, 46], and 75–87% using the AHI threshold of 10 e/h [42, 43]. In this respect, it is important to highlight that Ma et al. [42] obtained a substantially lower performance than the reported by Tsai et al. [43] using the ODI4. In addition, Van Eyck et al. [44] obtained a different diagnostic performance than Velasco et al. [45] and Brouillette et al. [38] prospectively validating the methods proposed in their studies. Apart from the different databases used in these works, this varying diagnostic performance could be accounted for by the fact that these works did not employ any validation strategy to further evaluate their methodological approaches.

In this research, we have compared the diagnostic performance of the developed signal processing methodologies with a classical oximetric index, ODI3, obtaining higher diagnostic capability in three independent and large cohorts of pediatric OSA patients. Furthermore, hold-out, bootstrapping, and K-fold cross validation strategies were used in the methodology of all the studies carried out in the doctoral thesis [1–3, 32], which, together with the large sample size, contribute to a higher generalization ability of our results.

In recent years, automated feature-engineering approaches have been used to enhance the diagnostic ability of the oximetry signal [5–8, 47–49]. These studies have employed signal processing and pattern recognition algorithms, also applying validation strategies to ensure the generalization of their results.

For an AHI cutoff of 1 e/h, the Acc obtained in these works ranged between 75.2% and 85.5% [5, 6, 8, 49]. In this research, accuracies were included in this range (77.6–82.7%) [3, 32]. Additionally, the studies showing higher Acc applied binary classifiers and used small databases (50 subjects in Álvarez et al. [5] and 176 subjects in Crespo et al. [6]), whereas in Vaquerizo et al. [3] and Vaquerizo et al. [32] we estimated the AHI of each patient using larger cohorts (see Table 5.2).

In the case of an AHI= 5 e/h, the diagnostic accuracy ranged 78.5–83.3% [5–8, 47, 49]. It is remarkable that our deep-learning model in Vaquerizo et al. [32] reported higher accuracies in the UofC (83.9%), CHAT (97.4%), and BUH (83.5%) databases. Similarly, our feature-engineering model in Vaquerizo et al. [2] also obtained a higher Acc (84.0%) in the UofC database, whereas the accuracies reported by our feature-engineering models in Vaquerizo et al. [1] and Vaquerizo et al. [2] were within this range (81.3% and 81.9%, respectively).

Finally, in the case of an AHI threshold of 10 e/h, Hornero et al. [8] and Xu et al. [49] reported 90.2% and 88.2% Acc, respectively. Vaquerizo et al. [1] achieved a lower accuracy for this cutoff (85.3%). However, our CNN-approach in Vaquerizo et al. [32] reported higher accuracies in the UofC (92.3%), CHAT (97.8%), and BUH (91.3%) datasets, whereas our feature-engineering proposal in Vaquerizo et al. [3] obtained 91.1% Acc in the UofC database.

Importantly, our feature-engineering and deep-learning approaches showed a high overall diagnostic ability in comparison with state-of-the-art studies, specially for the AHI cutoffs of 5 e/h and 10 e/h. The high diagnostic performance obtained with the proposed novel feature-engineering methodologies is consistent with the additional OSA-related information that these methods allow to quantify. Furthermore, the overall superior performance of our CNN-based methodology reinforces the ability of deep-learning approaches to learn complex features from oximetry dynamics related to apneic episodes in childhood OSA.

## 5.4 Limitations of the Study

The present doctoral thesis has shown the utility of novel feature-engineering and deep-learning approaches applied to the oximetry signal for diagnosing pediatric OSA and its severity. However, several limitations need to be considered.

One of the main limitations is that our proposals have not been assessed by population subgroups (i.e., age ranges, sex, and/or BMI groups among others), which hinders to discover in which OSA subgroups our oximetry-based approaches could be more appropriate, as well as to discern new phenotypes within pediatric OSA able to explain differences in the pathophysiology and severity of the disease.

Despite using three datasets involving a large number of pediatric patients, the sample size is not big enough to optimize and validate each methodology with subgroups of patients according to their clinical and physiological variables, which has resulted in a varying performance obtained with the CHAT, UofC, and BUH databases. Despite this sample size limitation, we used appropriate validation methodologies, so that optimization and validation groups from each database have similar clinical and sociodemographic characteristics.

Likewise, different recording devices and specific protocols were used for oximetry data collection in the three databases, which may influence the performance of the proposed methodologies. This is also a common problem in real-life clinical settings, as there exist multiple and pulse oximetry devices and polysomnography systems, even in the same sleep center. Notwithstanding, our proposals included a pre-processing stage to standardize the oximetry signals obtained from the different acquisition devices.

Another limitation relates to the use of the $SpO_2$ signal alone to automatically detect pediatric OSA. This limits the diagnostic ability of our proposed methodologies, since some apneic events are not linked to changes in oximetry dynamics [18, 50]. In addition, the total recording time was employed as a substitute for the total sleep time to estimate the AHI, as it is not possible to determine sleep stages from the oximetry signal alone. Nevertheless, our investigation has shown that a thorough analysis of the oximetry signal can reach a remarkable diagnostic performance.

Regarding the deep-learning methodology, we only used CNNs, which were originally designed for image analysis. Nonetheless, Ismail Fawaz et al. [51] reported that deep-learning architectures based on CNN are the most suitable for time series classification. Additionally, the CNN-based architecture was trained using only the CHAT dataset, as the UofC and BUH datasets do not contain annotations of the time location of apneic events, which may have contributed to a reduced performance in these datasets. However, our results showed that this methodology outperformed conventional methods in the three datasets. In spite of outperforming feature-engineering approaches, our CNN-based model also suffered from a lack of interpretation, which hinders to discover new knowledge regarding childhood OSA. Nonetheless, this 'black box' perception also exists in conventional pattern recognition algorithms.

Another limitation concerns to the use of the AHI as the reference measure for predicting the adverse outcomes of pediatric OSA. In this regard, recent investigations reported that novel measures of hypoxia obtained from the $SpO_2$ signal are better correlated with mortality, cardiovascular diseases, or cancer incidence than conventional respiratory indices (i.e., AHI or ODI) in adult OSA patients. Particularly, the nocturnal hypoxemia [52], the hypoxic load [53], the desaturation severity parameter [54, 55], and the hypoxic burden [56] have been proposed. As these measures have been found to further explain OSA consequences in adults, they could be also useful in the context of pediatric OSA.

The last limitation refers to the place where oximetry signals of the pediatric databases were obtained: supervised hospital facilities. In this respect, it would be interesting to further validate the methodologies proposed during the present doctoral thesis in a database of $SpO_2$ recordings acquired at children's home.

## References

1. Vaquerizo-Villar F, Álvarez D, Kheirandish-Gozal L, Gutiérrez-Tobal GC, Barroso-García V, Crespo A, del Campo F, Gozal D, Hornero R (2018) Utility of bispectrum in the screening of pediatric sleep apnea-hypopnea syndrome using oximetry recordings. Comput Meth Programs Biomed 156:141–149
2. Vaquerizo-Villar F, Álvarez D, Kheirandish-Gozal L, Gutiérrez-Tobal GC, Barroso-García V, Crespo A, del Campo F, Gozal D, Hornero R (2018) Wavelet analysis of oximetry recordings to assist in the automated detection of moderate-to-severe pediatric sleep apnea-hypopnea syndrome. PloS one 13(12):e0208502
3. Vaquerizo-Villar F, Álvarez D, Kheirandish-Gozal L, Gutiérrez-Tobal GC, Barroso-García V, Crespo A, Del Campo F, Gozal D, Hornero R (2018) Detrended fluctuation analysis of the oximetry signal to assist in paediatric sleep apnoea-hypopnoea syndrome diagnosis. Physiol Meas 39(11):114006
4. Alvarez D, Hornero R, Marcos JV, Wessel N, Penzel T, Glos M, Del Campo F (2013) Assessment of feature selection and classification approaches to enhance information from overnight oximetry in the context of apnea diagnosis. Int J Neural Syst 23(05):1350020
5. Álvarez D, Alonso-Álvarez ML, Gutiérrez-Tobal GC, Crespo A, Kheirandish-Gozal L, Gozal D, Terán-Santos J, Campo FD (2017) Automated screening of children with obstructive sleep apnea using nocturnal oximetry?: an alternative to respiratory polygraphy in unattended settings. J Clin Sleep Med 13(5):7–11
6. Crespo A, Álvarez D, Kheirandish-Gozal L, Gutiérrez-Tobal GC, Cerezo-Hernández A, Gozal D, Hornero R, Del Campo F (2018) Assessment of oximetry-based statistical classifiers as simplified screening tools in the management of childhood obstructive sleep apnea. Sleep Breathing 22(4):1063–1073

7. Garde A, Dehkordi P, Karlen W, Wensley D, Ansermino JM, Dumont GA (2014) Development of a screening tool for sleep disordered breathing in children using the phone oximeter. PloS one 9(11):e112959
8. Hornero R, Kheirandish-Gozal L, Gutiérrez-Tobal GC, Philby MF, Alonso-Álvarez ML, Álvarez D, Dayyat EA, Xu Z, Huang Y-S, Tamae Kakazu M, Li AM, Van Eyck A, Brockmann PE, Ehsan Z, Simakajornboon N, Kaditis AG, Vaquerizo-Villar F, Crespo Sedano A, Sans Capdevila O, von Lukowicz M, Terán-Santos J, Del Campo F, Poets CF, Ferreira R, Bertran K, Zhang Y, Schuen J, Verhulst S, Gozal D (2017) Nocturnal oximetry-based evaluation of habitually snoring children. Am J Respir Crit Care Med 196(12):1591–1598
9. Chua K, Chandran V, Acharya UR, Min C (2010) Application of higher order statistics/spectra in biomedical signals-a review. Med Eng Phys 32(7):679–689
10. Church GD (2012) The role of polysomnography in diagnosing and treating obstructive sleep apnea in pediatric patients. Curr Probl Pediatr Adolesc Health Care 42(1):22–25
11. Hunter SJ, Gozal D, Smith DL, Philby MF, Kaylegian J, Kheirandish-Gozal L (2016) Effect of sleep-disordered breathing severity on cognitive performance measures in a large community cohort of young school-aged children. Am J Respir Crit Care Med 194(6):739–747
12. Kaditis A, Kheirandish-Gozal L, Gozal D (2016) Pediatric OSAS: oximetry can provide answers when polysomnography is not available. Sleep Med Rev 27:96–105
13. Alonso-Álvarez ML, Canet T, Cubell-Alarco M, Estivill E, Fernández E, Gozal D, Jurado-Luque MJ, Lluch-Roselló MA, Martínez-Pérez F, Merino-Andren M, Pin-Arboledas G, Roure N, Sanmartí FX, Sans-Capdevila O, Segarra-Isern J (2011) Documento de consenso del síndrome de apneas-hipopneas durante el sueño en niños. Arch Bronconeumol 47(Supl 5):2–18
14. Atri R, Mohebbi M (2015) Obstructive sleep apnea detection using spectrum and bispectrum analysis of single-lead ecg signal. Physiol Meas 36(9):1963
15. Tagluk ME, Sezgin N (2011) A new approach for estimation of obstructive sleep apnea syndrome. Expert Syst Appl 38(5):5346–5351
16. Barroso-García V, Gutiérrez-Tobal GC, Kheirandish-Gozal L, Vaquerizo-Villar F, Álvarez D, del Campo F, Gozal D, Hornero R (2021) Bispectral analysis of overnight airflow to improve the pediatric sleep apnea diagnosis. Comput Biol Med 129:104167
17. Rioul O, Vetterli M (1991) Wavelets and signal processing. IEEE Sign Process Mag 8(4):14–38
18. Berry RB, Budhiraja R, Gottlieb DJ, Gozal D, Iber C, Kapur VK, Marcus CL, Mehra R, Parthasarathy S, Quan SF, Redline S, Strohl KP, Davidson Ward SL, Tangredi MM (2012) Rules for scoring respiratory events in sleep: update of the 2007 AASM manual for the scoring of sleep and associated events: deliberations of the sleep apnea definitions task force of the American Academy of Sleep Medicine. J Clin Sleep Med 8(5):597
19. Fontenla-Romero O, Guijarro-Berdinas B, Alonso-Betanzos A, Moret-Bonillo V (2005) A new method for sleep apnea classification using wavelets and feedforward neural networks. Artif Intell Med 34(1):65–76
20. Khandoker AH, Palaniswami M, Karmakar CK (2008) Support vector machines for automated recognition of obstructive sleep apnea syndrome from ecg recordings. IEEE Trans Inform Technol Biomed 13(1):37–48
21. Lin R, Lee R-G, Tseng C-L, Zhou H-K, Chao C-F, Jiang J-A (2006) A new approach for identifying sleep apnea syndrome using wavelet transform and neural networks. Biomed Eng Appl Basis Commun 18(03):138–143
22. Mendez M, Corthout J, Van Huffel S, Matteucci M, Penzel T, Cerutti S, Bianchi AM (2010) Automatic screening of obstructive sleep apnea from the ecg based on empirical mode decomposition and wavelet analysis. Physiol Measur 31(3):273
23. Lee Y, Bister M, Blanchfield P, Salleh Y (2004) Automated detection of obstructive apnea and hypopnea events from oxygen saturation signal. In: The 26th annual international conference of the IEEE engineering in medicine and biology society, vol. 1. IEEE, pp 321–324
24. del Campo F, Crespo A, Cerezo-Hernández A, Gutiérrez-Tobal GC, Hornero R, Álvarez D (2018) Oximetry use in obstructive sleep apnea. Expert Rev Respir Med 12(8):665–681
25. Hua C-C, Yu C-C (2017) Detrended fluctuation analysis of oxyhemoglobin saturation by pulse oximetry in sleep apnea syndrome. J Medical Biol Eng 37(6):791–799

26. Kaimakamis E, Tsara V, Bratsas C, Sichletidis L, Karvounis C, Maglaveras N (2016) Evaluation of a decision support system for obstructive sleep apnea with nonlinear analysis of respiratory signals. PloS one 11(3):e0150163
27. Lee J-M, Kim D-J, Kim I-Y, Park K-S, Kim SI (2002) Detrended fluctuation analysis of eeg in sleep apnea using mit/bih polysomnography data. Comput Biol Med 32(1):37–47
28. Penzel T, Kantelhardt J, Grote L, Peter J, Bunde A (2003) Comparison of detrended fluctuation analysis and spectral analysis for heart rate variability in sleep and sleep apnea. IEEE Trans Biomed Eng 50(10):1143–1151
29. Dehkordi P, Garde A, Karlen W, Petersen CL, Wensley D, Dumont GA, Ansermino JM (2016) Evaluation of cardiac modulation in children in response to apnea/hypopnea using the phone oximeter. Physiol Meas 37(2):187
30. Dehlink E, Tan H-L (2016) Update on paediatric obstructive sleep apnoea. J Thoracic Dis 8(2):224
31. LeCun Y, Bengio Y, Hinton G (2015) Deep learning. Nature 521(7553):436–444
32. Vaquerizo-Villar F, Alvarez D, Kheirandish-Gozal L, Gutierrez-Tobal GC, Barroso-Garcia V, Santamaria-Vazquez E, Del Campo F, Gozal D, Hornero R (2021) A convolutional neural network architecture to enhance oximetry ability to diagnose pediatric obstructive sleep apnea. IEEE J Biomed Health Inf 25(8):2906–2916
33. Biswal S, Sun H, Goparaju B, Brandon Westover M, Sun J, Bianchi MT (2018) Expert-level sleep scoring with deep neural networks. J Am Med Inf Assoc 25(12):1643–1650
34. Choi SH, Yoon H, Kim HS, Kim HB, Kwon HB, Oh SM, Lee YJ, Park KS (2018) Real-time apnea-hypopnea event detection during sleep by convolutional neural networks. Comput Biol Med 100(February):123–131
35. Nikkonen S, Afara IO, Leppänen T, Töyräs J (2019) Artificial neural network analysis of the oxygen saturation signal enables accurate diagnostics of sleep apnea. Sci Rep 9(1):13200
36. Van Steenkiste T, Groenendaal W, Deschrijver D, Dhaene T (2018) Automated sleep apnea detection in raw respiratory signals using long short-term memory neural networks. IEEE J Biomed Health Inf 23(6):2354–2364
37. Collop NA (2002) Scoring variability between polysomnography technologists in different sleep laboratories. Sleep Med 3(1):43–47
38. Brouillette RT, Morielli A, Leimanis A, Waters KA, Luciano R, Ducharme FM (2000) Nocturnal pulse oximetry as an abbreviated testing modality for pediatric obstructive sleep apnea. Pediatrics 105(2):405–412
39. Chang L, Wu J, Cao L (2013) Combination of symptoms and oxygen desaturation index in predicting childhood obstructive sleep apnea. Int J Pediatr Otorhinolaryngol 77(3):365–371
40. Nixon GM, Kermack AS, Davis GM, Manoukian JJ, Brown A, Brouillette RT (2004) Planning adenotonsillectomy in children with obstructive sleep apnea: the role of overnight oximetry. Pediatrics 113(1):e19–e25
41. Kirk VG, Bohn SG, Flemons WW, Remmers JE (2003) Comparison of home oximetry monitoring with laboratory polysomnography in children. Chest 124(5):1702–1708
42. Ma J-R, Huang J-J, Chen Q, Wu H-T, Xiao K-L, Zhang Y-T (2018) Value of pulse oximetry watch for diagnosing pediatric obstructive sleep apnea/hypopnea syndrome. Acta otolaryngologica 138(2):175–179
43. Tsai C-M, Kang C-H, Su M-C, Lin H-C, Huang E-Y, Chen C-C, Hung J-C, Niu C-K, Liao D-L, Yu H-R (2013) Usefulness of desaturation index for the assessment of obstructive sleep apnea syndrome in children. Int J Pediatr Otorhinolaryngol 77(8):1286–1290
44. Van Eyck A, Lambrechts C, Vanheeswijck L (2015) The role of nocturnal pulse oximetry in the screening for obstructive sleep apnea in obese children and adolescents. Sleep Med 16(11):1409–1412
45. Velasco CT, Suárez MD, Figueroa JM, Turienzo MD, Len FL, Mansilla E (2013) Pulse oximetry recording in children with adenotonsillar hypertrophy: usefulness in the diagnosis of obstructive sleep apnea syndrome. Arch Argentinos de Pediatr 111(3):196–201
46. Villa MP, Pietropaoli N, Supino MC, Vitelli O, Rabasco J, Evangelisti M, Del Pozzo M, Kaditis AG (2015) Diagnosis of pediatric obstructive sleep apnea syndrome in settings with limited resources. JAMA Otolaryngol-Head Neck Surg 141(11):990–996

47. Álvarez D, Crespo A, Vaquerizo-Villar F, Gutiérrez-Tobal GC, Cerezo-Hernández A, Barroso-García V, Ansermino JM, Dumont GA, Hornero R, Del Campo F, Garde A (2018) Symbolic dynamics to enhance diagnostic ability of portable oximetry from the phone oximeter in the detection of paediatric sleep apnoea. Phys Measur 39(10):104002
48. Crespo A, Álvarez D, Gutiérrez-Tobal GC, Vaquerizo-Villar F, Barroso-García V, Alonso-Álvarez ML, Terán-Santos J, Hornero R, del Campo F (2017) Multiscale entropy analysis of unattended oximetric recordings to assist in the screening of paediatric sleep apnoea at home. Entropy 19(6):284
49. Xu Z, Gutiérrez-Tobal GC, Wu Y, Kheirandish-Gozal L, Ni X, Hornero R, Gozal D (2019) Cloud algorithm-driven oximetry-based diagnosis of obstructive sleep apnoea in symptomatic habitually snoring children. Eur Respir J 53(2)
50. Marcus CL, Brooks LJ, Ward SD, Draper KA, Gozal D, Halbower AC, Jones J, Lehmann C, Schechter MS, Sheldon S, Shiffman RN, Spruyt K (2012) Diagnosis and management of childhood obstructive sleep apnea syndrome. Pediatrics 130(3):e714–e755
51. Ismail Fawaz H, Forestier G, Weber J, Idoumghar L, Muller PA (2019) Deep learning for time series classification: a review. Data Min Knowl Dis 33(4):917–963
52. Oldenburg O, Wellmann B, Buchholz A, Bitter T, Fox H, Thiem U, Horstkotte D, Wegscheider K (2016) Nocturnal hypoxaemia is associated with increased mortality in stable heart failure patients. Eur Heart J 37(21):1695–1703
53. Linz D, Colling S, Nußstein W, Debl K, Hohl M, Fellner C, Böhm M, Maier LS, Hamer OW, Buchner S et al (2018) Nocturnal hypoxemic burden is associated with epicardial fat volume in patients with acute myocardial infarction. Sleep Breathing 22(3):703–711
54. Kainulainen S, Töyräs J, Oksenberg A, Korkalainen H, Sefa S, Kulkas A, Leppänen T (2019) Severity of desaturations reflects osa-related daytime sleepiness better than ahi. J Clin Sleep Med 15(8):1135–1142
55. Kainulainen S, Duce B, Korkalainen H, Oksenberg A, Leino A, Arnardottir ES, Kulkas A, Myllymaa S, Töyräs J, Leppänen T (2020) Severe desaturations increase psychomotor vigilance task-based median reaction time and number of lapses in obstructive sleep apnoea patients. Eur Respir J 55(4)
56. Azarbarzin A, Sands SA, Stone KL, Taranto-Montemurro L, Messineo L, Terrill PI, Ancoli-Israel S, Ensrud K, Purcell S, White DP et al (2019) The hypoxic burden of sleep apnoea predicts cardiovascular disease-related mortality: the osteoporotic fractures in men study and the sleep heart health study. Eur Heart J 40(14):1149–1157

# Chapter 6
# Conclusions

This thesis is focused on the application of novel signal processing algorithms to improve the diagnostic capability of the oximetry signal in the simplification of pediatric OSA diagnosis. Feature-engineering and deep-learning methodologies were developed for this purpose. Among the feature-engineering approaches, it has been proposed three novel feature extraction algorithms (bispectrum, wavelet, and DFA) to provide additional OSA-related features from the $SpO_2$ signal in both the time and frequency domains. Additionally, a CNN-based deep-learning model was used to automatically extract all the relevant information from raw $SpO_2$ data linked to apneic events. Our results showed that the developed methodologies contribute to increase the diagnostic ability of overnight oximetry in the screening of childhood OSA.

In this chapter, the original contributions of this doctoral thesis are stated in Sect. 6.1. Then, the conclusions drawn from this doctoral thesis are indicated in Sect. 6.2. Finally, future research lines are listed in Sect. 6.3.

## 6.1 Contributions

The major contributions provided by this thesis are listed below:

(1) Novel automated feature-engineering and deep-learning models for the analysis of the $SpO_2$ signal, which have outperformed conventional approaches, thus enhancing the diagnostic capability of nocturnal oximetry in the framework of pediatric OSA [1–4].
(2) New oximetric indices through the application of bispectral, wavelet, and DFA, which have provided relevant and complimentary information on the changes in the oximetry dynamics associated to pediatric OSA and its severity. Although these methods had shown its usefulness to analyze physiological signals in adult



F. Vaquerizo Villar, *Automated Analysis of the Oximetry Signal to Simplify the Diagnosis of Pediatric Sleep Apnea*, Springer Theses,
https://doi.org/10.1007/978-3-031-32832-9_6

OSA patients, to the extent of our knowledge, this is the first time that these techniques are been applied in the context of childhood OSA [1–3].

(3) Optimum subsets obtained with the FCBF method, composed of conventional oximetry variables, as well as the new oximetric features derived from bispectral, wavelet, and DFA [1–3]. This highlights the relevancy and non-redundancy of the novel feature extraction methods (bispectral, wavelet, and DFA).
(4) High performance pattern recognition models focused on binary classification [2], multiclass classification [1], and regression [3]. These models were fed with the optimum subsets of OSA-related features and outperformed conventional approaches, as well as state-of-the-art approaches [1–3].
(5) Novel deep-learning model based on CNNs to automatically extract all the relevant information from the $SpO_2$ signal related to apneic events. This model was validated in a large sample of 3196 $SpO_2$ recordings from three independent datasets, showing a high diagnostic ability comparing with conventional feature-engineering methodologies and state-of-the-art studies. We believe that this is the first time that deep-learning algorithms are applied in the context of pediatric OSA diagnosis [4].
(6) Efficient screening protocols combining abbreviated test (oximetry) and artificial intelligence (pattern recognition and deep learning) able to minimize the number of PSGs. It was shown that these protocols would contribute to reduce the medical costs and waiting lists associated with the diagnosis of childhood OSA, as well as to reduce the children's discomfort during overnight PSG [2, 4].

## 6.2 Main Conclusions

The analysis and discussion of the obtained results (see Chaps. 4 and 5) lead to draw the principal conclusions of the present doctoral thesis, which are listed next:

(1) The proposed feature-engineering and deep-learning models outperform conventional features from the oximetry signal, as well as state-of-the-art approaches. Thus, the application of novel signal processing techniques allows to increase the diagnostic ability of the $SpO_2$ signal from nocturnal oximetry in the context of childhood OSA.
(2) Bispectrum can be used as a complementary tool to classical approaches in the characterization of OSA-related changes in children using $SpO_2$ recordings. Particularly, the changes in the bispectral amplitude associated to deviations of Gaussianity in the oximetry signal ($M1_{BISP}$) and the changes in the bispectral phase associated to a phase coupling between spectral components of the oximetry ($meanPa_{BISP}$) provide additional information to anthropometric parameters, ODI3, and PSD variables in the framework of childhood OSA.
(3) The DWT is a suitable tool to analyze the non-stationary properties, as well as the low frequency components occurring in the $SpO_2$ signal owing to pediatric OSA. Specifically, it was found that the concentration of the $D_9$ coefficients (0.0244–

0.0488 Hz) near zero ($M3_{D_9}$), the energy of the $D_9$ coefficients ($En_{D_9}$), and the changes of the energy distribution ($WE$) in the DWT profile of the oximetry signal provide complimentary information to conventional approaches.

(4) DFA is an appropriate tool to identify changes in the scaling behavior of the oximetry recordings related to pediatric OSA severity. Our findings suggest that the slope of the short-time scales of the DFA profile ($slope_1$) contains further information that contributes to further characterize OSA-related changes of the oximetry signal in children.

(5) From all the pattern recognition models, the SVM model fed with ODI3, statistical moments in the time domain, PSD and DWT-derived features has provided solid evidence to detect moderate-to-severe pediatric OSA (AHI $\geq$ 5 e/h), with an Acc of 84.0% and a LR+ of 14.6. This model can be used as a clinically valuable screening method for moderate-to-severe pediatric OSA patients.

(6) The CNN-based deep-learning model achieves a higher overall diagnostic performance than feature-engineering approaches in the framework of childhood OSA. Particularly, this model reached accuracies above 80% (97.4, 83.9, and 83.5%) for diagnosing moderate-to-severe-OSA and greater than 90% (97.8, 92.3, and 91.3%) for the detection of severe patients in the CHAT, UofC and BUH test sets, which also outperformed state-of-the-art studies. This is consistent with the improved predictive performance shown in recent years by deep-learning algorithms in a wide range of domains. Our findings suggest that deep learning could change the paradigm of biomedical data processing in the context of pediatric OSA.

(7) Deep-learning techniques show a high generalization ability, with a varying diagnostic performance that can be explained by differences in sampling rate, AHI distribution, age range, and patient characteristics among sleep datasets. Hence, clinical and sociodemographic parameters should be considered when validating our proposal in the clinical practice.

(8) The diagnostic protocol derived from our deep-learning model highlights the clinical applicability of overnight oximetry for the screening of childhood OSA. Particularly, the proposed screening tool would avoid the need for 45–70% (73.7, 50.0, and 45.9%) of complete PSGs in the CHAT, UofC, and BUH datasets. In this way, children would benefit from a more accessible and less intrusive diagnostic test based on the automated analysis of single-channel oximetry.

## 6.3 Future Research Lines

Several questions that arise from this investigation may be addressed in future work to complement our findings, and investigate other issues beyond the scope of this doctoral thesis. Next, the most interesting future research lines are listed:

(1) The evaluation of our automated signal processing methodologies in subgroups of children showing different clinical characteristics would help to character-

ize the physiological patterns shared by the OSA pediatric population, as well as to identify those phenotypes within pediatric OSA where oximetry-based approaches achieve a higher performance.

(2) The field of deep learning is living breakthrough advances thanks to the development of novel deep neural architectures, such as attention or inception networks. In this respect, the application of more advanced deep-learning architectures is another interesting future line of investigation that may contribute to improve the diagnostic performance of oximetry-based approaches.

(3) The general performance of the oximetry signal may also be increased by using pretrained deep-learning networks designed for time series classification, similar to the pretrained deep-learning architectures existing in the field of image processing.

(4) Another interesting future research could be the application of eXplainable Artificial Intelligence techniques to detect new patterns/attributes inherent to the oximetry signal linked with the severity of pediatric OSA.

(5) The evaluation of the proposed methodologies in different types of pulse oximeters and recording systems would help to know if the diagnostic performance is affected by the technical features of the recording equipment, as well as to improve the pre-processing stage.

(6) The acquisition of the PPG signal with the pulse oximetry sensor would also be interesting, since the PPG signal contains information of the changes in the autonomic nervous system and respiratory activity related to sleep stages and apneic events. In this way, the PPG signal may help to improve the diagnostic ability of our proposal.

(7) One natural way to continue our research would be to validate the proposed methodology in oximetry recordings acquired with portable devices at children's home, as the final goal is to perform at-home screening tests for pediatric OSA based on nocturnal oximetry.

(8) Another future objective would be to assess the correlation of novel hypoxic measures with cardiovascular, metabolic, behavioral, and neurocognitive variables in pediatric OSA patients, as well as to propose new estimates of the level of hypoxia.

## References

1. Vaquerizo-Villar F, Álvarez D, Kheirandish-Gozal L, Gutiérrez-Tobal GC, Barroso-García V, Crespo A, del Campo F, Gozal D, Hornero R (2018) Utility of bispectrum in the screening of pediatric sleep apnea-hypopnea syndrome using oximetry recordings. Comput Methods Programs Biomed 156:141–149
2. Vaquerizo-Villar F, Álvarez D, Kheirandish-Gozal L, Gutiérrez-Tobal GC, Barroso-García V, Crespo A, del Campo F, Gozal D, Hornero R (2018) Wavelet analysis of oximetry recordings to assist in the automated detection of moderate-to-severe pediatric sleep apnea-hypopnea syndrome. PloS One 13(12):e0208502

3. Vaquerizo-Villar F, Álvarez D, Kheirandish-Gozal L, Gutiérrez-Tobal GC, Barroso-García V, Crespo A, Del Campo F, Gozal D, Hornero R (2018) Detrended fluctuation analysis of the oximetry signal to assist in paediatric sleep apnoea-hypopnoea syndrome diagnosis. Physiol Meas 39(11):114006
4. Vaquerizo-Villar F, Alvarez D, Kheirandish-Gozal L, Gutierrez-Tobal GC, Barroso-Garcia V, Santamaria-Vazquez E, Del Campo F, Gozal D, Hornero R (2021) A convolutional neural network architecture to enhance oximetry ability to diagnose pediatric obstructive sleep apnea. IEEE J Biomed Health Inf 25(8):2906–2916

# Appendix A
# About the Author

Fernando Vaquerizo Villar is a postdoctoral researcher of the Biomedical Engineering Group at University of Valladolid. He received the bachelor degree in Specific Telecommunication Technologies Engineering from the University of Valladolid in 2014, the master degree in Telecommunication Engineering in 2016, and the PhD degree in the Ph.D. Program in Information and Telecommunication Technologies in 2021. His research is primarily focused on the automated analysis of biomedical signals during sleep using signal processing and artificial intelligence algorithms, mainly in the field of sleep medicine. As a consequence of his research activity, he has published published 21 articles up to now in international journals indexed in the JCR, 4 book chapters, and 46 communications in national (26) and international (20) congresses. The impact of his research has been recognized with various national and international awards. In this sense, his doctoral thesis has received the Springer Thesis Award of the Spanish Committee of Automatics in the field of Bioengineering, which has led to publish his thesis in Springer Series. In addition, he has been awarded for JCR articles (Young Researchers Award for the best articles published in the years 2020 and 2021 in the Bioengineering area of the *Centro de Investigación Biomédica en Red de Bioingeniería, Biomateriales y Nanomedicina*). He has also obtained awards for various contributions in congresses (SES 2017



F. Vaquerizo Villar, *Automated Analysis of the Oximetry Signal to Simplify the Diagnosis of Pediatric Sleep Apnea*, Springer Theses,
https://doi.org/10.1007/978-3-031-32832-9

and SEPAR 2018), prizes for research in regional competitive competition projects (SOCALPAR 2016 and SOCALPAR 2019), an award granted by the Castilla y León Health Cluster (BIOTECYL), and the Extraordinary End of Career Award for the Specific Telecommunication Technologies Engineering Degree at the University of Valladolid.

MIX
Papier aus verantwortungsvollen Quellen
Paper from responsible sources
FSC® C105338

If you have any concerns about our products,
you can contact us on
**ProductSafety@springernature.com**

In case Publisher is established outside the EU,
the EU authorized representative is:
**Springer Nature Customer Service Center GmbH**
**Europaplatz 3, 69115 Heidelberg, Germany**

Printed by Libri Plureos GmbH
in Hamburg, Germany